AF678103

# Lesion Detection and Analysis Using Optical Imaging

# Lesion Detection and Analysis Using Optical Imaging

Editor

**Viktor Dremin**

MDPI • Basel • Beijing • Wuhan • Barcelona • Belgrade • Manchester • Tokyo • Cluj • Tianjin

MDPI

*Editor*
Viktor Dremin
Aston University
Birmingham, UK

*Editorial Office*
MDPI
St. Alban-Anlage 66
4052 Basel, Switzerland

This is a reprint of articles from the Special Issue published online in the open access journal *Diagnostics* (ISSN 2075-4418) (available at: https://www.mdpi.com/journal/diagnostics/special_issues/Lesion_Optical_Imaging).

For citation purposes, cite each article independently as indicated on the article page online and as indicated below:

LastName, A.A.; LastName, B.B.; LastName, C.C. Article Title. *Journal Name* **Year**, *Volume Number*, Page Range.

**ISBN 978-3-0365-7990-0 (Hbk)**
**ISBN 978-3-0365-7991-7 (PDF)**

© 2023 by the authors. Articles in this book are Open Access and distributed under the Creative Commons Attribution (CC BY) license, which allows users to download, copy and build upon published articles, as long as the author and publisher are properly credited, which ensures maximum dissemination and a wider impact of our publications.
The book as a whole is distributed by MDPI under the terms and conditions of the Creative Commons license CC BY-NC-ND.

# Contents

# About the Editor

**Viktor Dremin**

Dr. Viktor Dremin received his MSc and PhD degrees from Orel State University, Orel, Russia, in 2013 and 2017, respectively. During the first several years of his career, he was involved in the development of new scientific devices for optical remote sensing of the Earth and terrestrial planets as an optoelectronics design engineer of the company "Astron Electronics" (Russia). As a PhD student, Dr. Dremin also conducted active research in biomedical engineering and biophotonics at the R&D Center of Biomedical Photonics (Orel State University, Russia). Part of the PhD research was carried out at Aston University (UK) via Erasmus+ programme. After his PhD viva (2017), his work on the development of an imaging system for skin chromophores visualisation was supported by a EDUFI Fellowship (Finland). In 2019, his project to develop a multimodal hyperspectral system for the diagnosis of the glycation of biological tissues received the support of a prestigious postdoctoral grant awarded by the Marie Skłodowska-Curie Individual Fellowships programme. His current research interests include biomedical imaging, optical tools for assessment of metabolic activity of biological tissues, modelling of optical radiation propagation in biological tissues, etc. He has various prestigious national and international awards. He is the PI and Co-Investigator of a number of projects supported by various international foundations (H2020, RSF, RFBR, Academy of Finland, etc.). He has authored and co-authored over 100 articles in refereed journals and conference proceedings, 9 patents, and 4 book chapters.

diagnostics

*Editorial*

# Lesion Detection and Analysis Using Optical Imaging

**Viktor Dremin** [1,2]

[1] Research & Development Center of Biomedical Photonics, Orel State University, 302026 Orel, Russia; v.dremin1@aston.ac.uk

[2] College of Engineering and Physical Sciences, Aston University, Birmingham B4 7ET, UK

The biomedical application of optical spectroscopy and imaging is currently an active, developing area of research, supported by recent technical progress in the development of light sources and detectors. Furthermore, the development of portable and low-cost optical imaging systems shows strong potential for the implementation of these technologies in everyday life and the daily practice of clinicians. In recent years, many studies have demonstrated that the use of modern optical imaging methods in conjunction with a priori data, as well as advanced data mining approaches, can significantly improve the quality of the medical diagnostic services provided [1,2].

The probing radiation of optical diagnostic methods is scattered due to random spatial variations in tissue morphology. Detection of diffuse reflected or transmitted light can provide information on the scattering and absorbing components of biological tissue. Changes in tissue morphology, including hyperplasia, collagen degradation in the extracellular matrix, and an increase in the nuclear–cytoplasmic ratio associated with the progression of various diseases, can affect the scattering signals. In addition, changes in hemoglobin-absorbing properties can assist with angiogenesis processes, the presence of tissue hypoxia and ischemia, etc. At the same time, incident radiation can excite biological tissue molecules, causing fluorescent radiation. Most endogenous fluorophores are associated with the morphology of the tissue or with various metabolic processes responsible for the functional state of the tissue. Mitochondrial function is an important parameter of tissue viability. According to indicators of respiratory chain activity, it is possible to predict cell death, diagnose tissue ischemia, or, on the contrary, talk about its malignant activity. Additionally, the dynamic change of scattering centers of biological tissues probed by laser radiation leads to the formation of speckle fields, the study of which makes it possible to obtain perfusion (blood flow) maps in various anatomical areas. Thus, biophotonic methods can provide unique opportunities for structural and functional analysis of biological tissues, as well as for early and non-invasive diagnosis and monitoring of the effectiveness of therapy in various diseases [1].

Biophotonics has broad prospects for development, since relatively cheap optical technologies allow images to be obtained and human tissues and organs to be influenced in real time with micron resolution and without the use of ionizing radiation. These technologies have found numerous applications in scientific research and clinical practice: dynamic light scattering methods (laser Doppler flowmetry [3,4], laser speckle contrast imaging [5,6], diffusing wave spectroscopy [7]); diffuse reflectance spectroscopy and visualization, including hyperspectral imaging [8,9]; fluorescence spectroscopy and visualization including lifetime measurements [10,11]; polarimetry [12,13]; videocapillaroscopy [14,15]; optical coherence tomography [16]; THz spectroscopy [17]; various microscopy techniques [18], etc. Modern laboratory and clinical studies show that the use of these instrumental approaches based on the registration of optical irradiation makes it possible to study various diseases from the cellular to the organismal level. Many of the mentioned methods have already reliably occupied their niches in everyday biomedical practice, such as pulse oximetry, optical coherence tomography, indocyanine green fluorescent imaging, and various microscopy realizations. However, further improvements in various optical technologies and their application methodology are still required prior to their large-scale implementation.

**Citation:** Dremin, V. Lesion Detection and Analysis Using Optical Imaging. *Diagnostics* **2023**, *13*, 1565. https://doi.org/10.3390/diagnostics13091565

Received: 6 April 2023
Accepted: 25 April 2023
Published: 27 April 2023

**Copyright:** © 2023 by the author. Licensee MDPI, Basel, Switzerland. This article is an open access article distributed under the terms and conditions of the Creative Commons Attribution (CC BY) license (https://creativecommons.org/licenses/by/4.0/).

This Special Issue is one of many small steps on this long and fascinating journey. It highlights the advantages and unique features of using optical diagnostic technologies for the detection and analysis of lesions. The articles discuss the development and application of optical methods, expanding the current diagnostic capabilities of various diseases, and other aspects of spectroscopy and imaging in biology and clinical practice.

In a case study [19], Y. Goh et al. demonstrated the potential of photoacoustic imaging (PA) as a supplement to ultrasound (US) to increase the reliability of lipoma diagnosis. In this case, the authors showed that, using PA, it is possible to study the biochemical characteristics of fat necrosis. PA can identify the "mass-like" hypoechoic areas on US images as fat-containing, rather than fat-replacing. PA provided information on liquefied necrotic fat from cystic degeneration and fibrosis around the cavity. Therefore, the new biochemical information on fat and collagen content provided by PA can help to recognize ambiguous US results and increase confidence in the diagnosis of fat necrosis.

Several studies in this Special Issue concentrate on the application of Raman spectroscopy in combination with other techniques. Y. Khristoforova et al. combined patient data with Raman and autofluorescence spectral characteristics to better identify skin tumors (melanoma, basal cell carcinoma, squamous cell carcinoma and different benign tumors) [20]. The spectral characteristics of the tumors and patient data were used to construct classifiers based on projection on latent structures and discriminant analysis. Gender, age, and tumor location were found to be significant to the classification of malignant versus benign neoplasms. S. Al-Shareefi et al. applied Raman spectroscopy in dentistry and demonstrated that this method can detect changes in the content of minerals and the collagen matrix in various carious areas [21]. The caries-infected dentin showed a low phosphate peak and higher amide I, III and C-H bond peaks. The amide peaks (I, III) varied in occlusal lesions, as opposed to proximal. Additionally, there was a significant correlation between the mineral:matrix peak ratio and the equivalent Vickers microhardness number within carious lesions.

J. Sachs et al. evaluated the possible impact of dual-energy computed tomography (DECT) on cataract formation [22]. The authors studied the attenuation of X-ray radiation via the crystalline lens, considering clinical and demographic data. The authors concluded that the lens of women and people of color had a higher attenuation during DECT, which may indicate a higher density or increased calcium concentration and a high probability of cataract formation.

P. Glazkova et al. studied the possibilities of incoherent optical fluctuation flowmetry (IOFF) to analyze the perfusion of foot tissues in patients with diabetes mellitus [23]. Perfusion in foot tissues was also assessed using transcutaneous oxygen pressure measurements ($TcPO_2$) as a reference standard. The high correlation coefficients were shown between the measurements of the perfusion parameters using IOFF and $TcPO_2$. This study demonstrates that the IOFF method allows the identification of patients with a critical decrease in $TcPO_2 < 20$ mmHg with sensitivity and specificity of 85.7% and 90.0%, respectively.

The work of E. Zharkikh et al. focuses on studying microcirculation features in patients who underwent COVID-19 using novel wearable laser Doppler flowmetry (LDF) devices [24]. In patients undergoing rehabilitation after contracting COVID-19, a decrease in skin perfusion and changes in the amplitude–frequency behavior of the LDF signal were found. The data obtained confirmed the presence of microcirculatory dysfunction for a long period after recovery from COVID-19.

In a study by E. Cinotti et al., the use of line-field confocal optical coherence tomography (LC-OCT) for the diagnosis of skin cancer was demonstrated [25]. LC-OCT is a new non-invasive imaging technology that combines the benefits of optical coherence tomography and reflectance confocal microscopy in terms of spatial resolution, penetration depth and image orientation, overcoming their corresponding disadvantages and limitations. Considering different types of skin malignant tumors (basal cell carcinoma, squamous cell carcinoma, melanoma), a statistically significant increase in specificity was obtained from 0.73 for dermoscopy to 0.87 for LC-OCT. At the same time, the sensitivity was similar for the two imaging methods (0.95). The increased specificity was mainly due to LC-OCT's ability to differentiate basal cell carcinoma from other malignancies.

Two studies were devoted to the use of diffuse reflectance spectroscopy (DRS). V. Perekatova et al. reported on the comparative study of self-calibrating and single-slope DRS in terms of stability against various measurement disturbances [26]. The authors designed an experimental setup for DRS in a wide spectral range (from VIS to NIR), including a fiber-optic probe with two fibers for delivery and two used to receive light. This setup was capable of measuring with both single- and dual-slope (self-calibrating) approaches. The resistance of self-calibrating and traditional single-slope approaches to different instrumental disturbances was studied in phantom and *in vivo* experiments in human skin. The new method showed high stability to instrumental disturbances introduced into the source and detection channels, while the traditional single-slope approach demonstrated stability only to perturbations in the source channels.

In another study [27] A. Selifonov et al. used spectrophotometry measurements of diffuse reflectance and total transmittance in a wide spectral range (from 200 to 800 nm) to analyze the optical and molecular diffusion properties of cat ovarian tissues in the follicular and luteal phases using glycerol as an optical clearing agent. The authors found that the efficiency of optical clearing was significantly lower for the ovaries in the luteal phase compared with the follicular phase. The authors claim that the ability to recognize the phase in which the ovaries are stated, using their approach, could be useful in the context of cryopreservation, new reproductive technologies and ovarian implantation.

Finally, I. Zlotnikov et al. demonstrated the use of confocal laser scanning microscopy (CLSM) and Fourier-transform infrared spectroscopy (FTIR) to study the interactions of antibacterial drugs with bacterial cells [28]. More specifically, using these optical techniques, the authors studied the efflux effect of drug molecules from E.coli bacterial cells. It was demonstrated that eugenol, which acts as an adjuvant to rifampicin, significantly increased the penetration of the antibiotic and the maintenance of its intracellular concentration. In addition, optical methods were used to study bacteria localized inside macrophages, where the availability of bacteria for antibiotics is reduced, and an approach for effective drug delivery inside macrophages was demonstrated.

Thus, this Special Issue contains new developments and advanced ideas pertaining to optical spectroscopy and imaging for active translation to biological and clinical practice. The published manuscripts provide new insight into state-of-the-art modern methods of biomedical optics and expand readers' knowledge of the possible areas of their application.

The peculiarity of biophotonics is that it combines physics, biology and medicine. The interdisciplinarity of biophotonics allows scientists from different fields to unite and collaborate to solve complex problems. Biophotonics and related fields are continuously developing and will undoubtedly become the main avenue for the development of biomedical engineering in the coming decades.

**Funding:** The work was supported by the Russian Science Foundation under the project No. 22-75-10088.

**Conflicts of Interest:** The author declares no conflict of interest.

## References


1. Tuchin, V.V. *Handbook of Optical Biomedical Diagnostics, Volume 2: Methods*, 2nd ed.; SPIE: Bellingham, DC, USA, 2016; p. 688.
2. Zharkikh, E.; Dremin, V.; Zherebtsov, E.; Dunaev, A.; Meglinski, I. Biophotonics methods for functional monitoring of complications of diabetes mellitus. *J. Biophotonics* **2020**, *13*, e202000203. [CrossRef] [PubMed]
3. Dremin, V.; Zherebtsov, E.; Makovik, I.; Kozlov, I.; Sidorov, V.; Krupatkin, A.; Dunaev, A.; Rafailov, I.; Litvinova, K.; Sokolovski, S.; et al. Laser Doppler flowmetry in blood and lymph monitoring, technical aspects and analysis. *Proc. SPIE* **2017**, *10063*, 1006303.
4. Zherebtsov, E.; Kozlov, I.; Dremin, V.; Bykov, A.; Dunaev, A.; Meglinski, I. Diagnosis of skin vascular complications revealed by time-frequency analysis and laser Doppler spectrum decomposition. *IEEE Trans. Biomed. Eng.* **2023**, *70*, 3–14. [CrossRef]
5. Mizeva, I.; Dremin, V.; Potapova, E.; Zherebtsov, E.; Kozlov, I.; Dunaev, A. Wavelet Analysis of the Temporal Dynamics of the Laser Speckle Contrast in Human Skin. *IEEE Trans. Biomed. Eng.* **2019**, *67*, 1882–1889. [CrossRef] [PubMed]
6. Vaz, P.G.; Humeau-Heurtier, A.; Figueiras, E.; Correia, C.; Cardoso, J. Laser Speckle Imaging to Monitor Microvascular Blood Flow: A Review. *IEEE Rev. Biomed. Eng.* **2016**, *9*, 106–120. [CrossRef]

7. Meglinski, I.; Tuchin, V.V. Diffusing wave spectroscopy: Application for blood diagnostics. In *Handbook of Coherent-Domain Optical Methods: Biomedical Diagnostics, Environmental Monitoring, and Materials Science*; Springer: New York, NY, USA, 2013; pp. 149–166.
8. Zherebtsov, E.; Dremin, V.; Popov, A.; Doronin, A.; Kurakina, D.; Kirillin, M.; Meglinski, I.; Bykov, A. Hyperspectral imaging of human skin aided by artificial neural networks. *Biomed. Opt. Express* **2019**, *10*, 3545. [CrossRef] [PubMed]
9. Dremin, V.V.; Zherebtsov, E.A.; Popov, A.P.; Meglinski, I.V.; Bykov, A.V. Hyperspectral imaging of diabetes mellitus skin complications. In *Biomedical Photonics for Diabetes Research*; CRC Press: Boca Raton, FL, USA, 2022; pp. 177–195.
10. Mycek, M.A.; Pogue, B.W. *Handbook of Biomedical Fluorescence*; CRC Press: Boca Raton, FL, USA, 2003; p. 694.
11. Zherebtsov, E.; Potapova, E.; Mamoshin, A.; Shupletsov, V.; Kandurova, K.; Dremin, V.; Abramov, A.; Dunaev, A. Fluorescence lifetime needle optical biopsy discriminates hepatocellular carcinoma. *Biomed. Opt. Express* **2022**, *13*, 633–646. [CrossRef]
12. He, C.; He, H.; Chang, J.; Chen, B.; Ma, H.; Booth, M.J. Polarisation optics for biomedical and clinical applications: A review. *Light Sci. Appl.* **2021**, *10*, 194. [CrossRef]
13. Ivanov, D.; Dremin, V.; Borisova, E.; Bykov, A.; Novikova, T.; Meglinski, I.; Ossikovski, R. Polarization and depolarization metrics as optical markers in support to histopathology of *ex vivo* colon tissue. *Biomed. Opt. Express* **2021**, *12*, 4560–4572. [CrossRef]
14. Dremin, V.; Kozlov, I.; Volkov, M.; Margaryants, N.; Potemkin, A.; Zherebtsov, E.; Dunaev, A.; Gurov, I. Dynamic evaluation of blood flow microcirculation by combined use of the laser Doppler flowmetry and high-speed videocapillaroscopy methods. *J. Biophotonics* **2019**, *12*, e201800317. [CrossRef]
15. Margaryants, N.B.; Sidorov, I.S.; Volkov, M.V.; Gurov, I.P.; Mamontov, O.V.; Kamshilin, A.A. Visualization of skin capillaries with moving red blood cells in arbitrary area of the body. *Biomed. Opt. Express* **2019**, *10*, 4896–4906. [CrossRef] [PubMed]
16. de Boer, J.F.; Leitgeb, R.; Wojtkowski, M. Twenty-five years of optical coherence tomography: The paradigm shift in sensitivity and speed provided by Fourier domain OCT [Invited]. *Biomed. Opt. Express* **2017**, *8*, 3248–3280. [CrossRef] [PubMed]
17. Smolyanskaya, O.; Chernomyrdin, N.; Konovko, A.; Zaytsev, K.; Ozheredov, I.; Cherkasova, O.; Nazarov, M.; Guillet, J.P.; Kozlov, S.; Kistenev, Y.V.; et al. Terahertz biophotonics as a tool for studies of dielectric and spectral properties of biological tissues and liquids. *Prog. Quantum Electron.* **2018**, *62*, 1–77. [CrossRef]
18. Thorn, K. A quick guide to light microscopy in cell biology. *Mol. Biol. Cell.* **2016**, *27*, 219–222. [CrossRef] [PubMed]
19. Goh, Y.; Balasundaram, G.; Tan, H.M.; Putti, T.C.; Ng, C.W.Q.; Fang, E.; Bi, R.; Tang, S.W.; Buhari, S.A.; Hartman, M.; et al. Photoacoustic Tomography Appearance of Fat Necrosis: A First-in-Human Demonstration of Biochemical Signatures along with Histological Correlation. *Diagnostics* **2022**, *12*, 2456. [CrossRef]
20. Khristoforova, Y.; Bratchenko, I.; Bratchenko, L.; Moryatov, A.; Kozlov, S.; Kaganov, O.; Zakharov, V. Combination of Optical Biopsy with Patient Data for Improvement of Skin Tumor Identification. *Diagnostics* **2022**, *12*, 2503. [CrossRef]
21. Al-Shareefi, S.; Addie, A.; Al-Taee, L. Biochemical and Mechanical Analysis of Occlusal and Proximal Carious Lesions. *Diagnostics* **2022**, *12*, 2944. [CrossRef]
22. Sachs, J.R.; Nahmias, J.A.; Hiatt, K.D.; Bomar, J.G.; West, T.G.; Bunch, P.M.; Benayoun, M.D.; Lack, C.; Thompson, A.C. Study of the Natural Crystalline Lens Characteristics Using Dual-Energy Computed Tomography. *Diagnostics* **2022**, *12*, 2857. [CrossRef]
23. Glazkova, P.; Glazkov, A.; Kulikov, D.; Zagarov, S.; Kovaleva, Y.; Babenko, A.; Kononova, Y.; Kitaeva, E.; Britvin, T.; Mazur, N.; et al. Incoherent Optical Fluctuation Flowmetry: A New Method for the Assessment of Foot Perfusion in Patients with Diabetes-Related Lower-Extremity Complications. *Diagnostics* **2022**, *12*, 2922. [CrossRef]
24. Zharkikh, E.V.; Loktionova, Y.I.; Fedorovich, A.A.; Gorshkov, A.Y.; Dunaev, A.V. Assessment of Blood Microcirculation Changes after COVID-19 Using Wearable Laser Doppler Flowmetry. *Diagnostics* **2023**, *13*, 920. [CrossRef]
25. Cinotti, E.; Brunetti, T.; Cartocci, A.; Tognetti, L.; Suppa, M.; Malvehy, J.; Perez-Anker, J.; Puig, S.; Perrot, J.L.; Rubegni, P. Diagnostic Accuracy of Line-Field Confocal Optical Coherence Tomography for the Diagnosis of Skin Carcinomas. *Diagnostics* **2023**, *13*, 361. [CrossRef] [PubMed]
26. Perekatova, V.; Kostyuk, A.; Kirillin, M.; Sergeeva, E.; Kurakina, D.; Shemagina, O.; Orlova, A.; Khilov, A.; Turchin, I. VIS-NIR Diffuse Reflectance Spectroscopy System with Self-Calibrating Fiber-Optic Probe: Study of Perturbation Resistance. *Diagnostics* **2023**, *13*, 457. [CrossRef] [PubMed]
27. Selifonov, A.A.; Rykhlov, A.S.; Tuchin, V.V. The Glycerol-Induced Perfusion-Kinetics of the Cat Ovaries in the Follicular and Luteal Phases of the Cycle. *Diagnostics* **2023**, *13*, 490. [CrossRef] [PubMed]
28. Zlotnikov, I.D.; Ezhov, A.A.; Vigovskiy, M.A.; Grigorieva, O.A.; Dyachkova, U.D.; Belogurova, N.G.; Kudryashova, E.V. Application Prospects of FTIR Spectroscopy and CLSM to Monitor the Drugs Interaction with Bacteria Cells Localized in Macrophages for Diagnosis and Treatment Control of Respiratory Diseases. *Diagnostics* **2023**, *13*, 698. [CrossRef] [PubMed]


**Disclaimer/Publisher's Note:** The statements, opinions and data contained in all publications are solely those of the individual author(s) and contributor(s) and not of MDPI and/or the editor(s). MDPI and/or the editor(s) disclaim responsibility for any injury to people or property resulting from any ideas, methods, instructions or products referred to in the content.

diagnostics

MDPI

*Interesting Images*

# Photoacoustic Tomography Appearance of Fat Necrosis: A First-in-Human Demonstration of Biochemical Signatures along with Histological Correlation

**Yonggeng Goh [1,†], Ghayathri Balasundaram [2,†], Hui Min Tan [3], Thomas Choudary Putti [3], Celene Wei Qi Ng [4], Eric Fang [1], Renzhe Bi [2], Siau Wei Tang [4], Shaik Ahmad Buhari [4], Mikael Hartman [4], Ching Wan Chan [4], Yi Ting Lim [1], Malini Olivo [2,*] and Swee Tian Quek [1,*]**

1 Department of Diagnostic Imaging, National University Hospital, 5 Lower Kent Ridge Road, Singapore 119074, Singapore
2 Translational Biophotonics Laboratory, Institute of Bioengineering & Bioimaging, Agency for Science, Technology & Research, 11 Biopolis Way, #02-02 Helios, Singapore 138667, Singapore
3 Department of Pathology, National University Hospital, 5 Lower Kent Ridge Road, Singapore 119074, Singapore
4 Department of Breast Surgery, National University Hospital, 5 Lower Kent Ridge Road, Singapore 119074, Singapore
* Correspondence: malini_olivo@ibb.a-star.edu.sg (M.O.); swee_tian_quek@nuhs.edu.sg (S.T.Q.); Tel.: +65-6478 7018 (M.O.)
† These authors contributed equally to this work.

**Citation:** Goh, Y.; Balasundaram, G.; Tan, H.M.; Putti, T.C.; Ng, C.W.Q.; Fang, E.; Bi, R.; Tang, S.W.; Buhari, S.A.; Hartman, M.; et al. Photoacoustic Tomography Appearance of Fat Necrosis: A First-in-Human Demonstration of Biochemical Signatures along with Histological Correlation. *Diagnostics* **2022**, *12*, 2456. https://doi.org/10.3390/diagnostics12102456

Academic Editor: Viktor Dremin

Received: 20 September 2022
Accepted: 10 October 2022
Published: 11 October 2022

**Publisher's Note:** MDPI stays neutral with regard to jurisdictional claims in published maps and institutional affiliations.

**Copyright:** © 2022 by the authors. Licensee MDPI, Basel, Switzerland. This article is an open access article distributed under the terms and conditions of the Creative Commons Attribution (CC BY) license (https://creativecommons.org/licenses/by/4.0/).

**Abstract:** A 50-year-old woman with no past medical history presented with a left anterior chest wall mass that was clinically soft, mobile, and non-tender. A targeted ultrasound (US) showed findings suggestive of a lipoma. However, focal "mass-like" nodules seen within the inferior portion suggested malignant transformation of a lipomatous lesion called for cross sectional imaging, such as MRI or invasive biopsy or excision for histological confirmation. A T1-weighted image demonstrated a large lipoma that has a central fat-containing region surrounded by an irregular hypointense rim in the inferior portion, confirming the benignity of the lipoma. An ultrasound-guided photoacoustic imaging (PA) of the excised specimen to derive the biochemical distribution demonstrated the "mass-like" hypoechoic regions on US as fat-containing, suggestive of benignity of lesion, rather than fat-replacing suggestive of malignancy. The case showed the potential of PA as an adjunct to US in improving the diagnostic confidence in lesion characterization.

**Keywords:** photoacoustic; lipoma; ultrasound; MRI

Fat necrosis, or cell death of adipose tissue, is a common benign condition that occurs from the lack of oxygen supply to adipose tissue [1]. As common causes include trauma or post-surgical changes [2], fat necrosis often presents as a palpable soft tissue mass at superficial regions [3].

Ultrasound (US) is the first-line imaging tool for these superficial lesions, but imaging appearances are extremely varied [4] due to the age of the lesion, which manifests as varying degrees of hardening, fibrosis, and degeneration. This often results in a diagnostic dilemma, which necessitates further cross-sectional imaging or invasive procedures, such as biopsy or excision for histological confirmation (Figures 1 and 2). There is, hence, an unmet clinical need for an adjunct imaging modality to US to improve diagnostic capability.

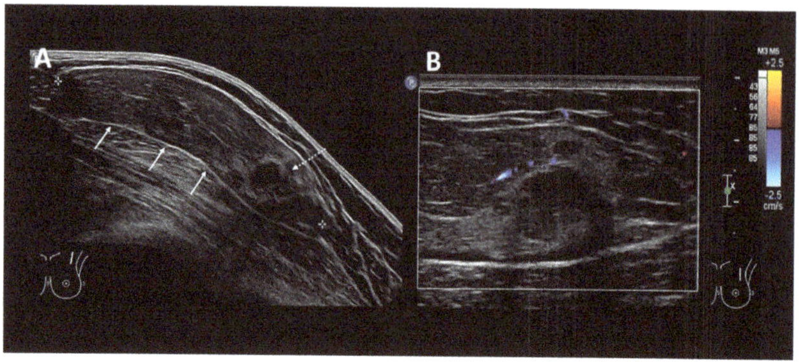


**Figure 1.** A 50-year-old woman with no past medical history presented with a left anterior chest wall mass. Clinically, the mass was soft, mobile, and non-tender. (**A**): US image of the left anterior chest wall mass shows a wider than tall mass with well-circumscribed margins (bold arrows). It contains internal fat echogenicity and is in keeping with a fat-containing lesion (i.e., lipoma). Within the inferior portions (dotted arow), there are focal "mass-like" hypoechoic nodules seen, which are surrounded by a hyperechoic capsule. (**B**): Doppler US of the focal "mass-like" nodules performed, demonstrated mild increased peripheral vascularity around the hyperechoic capsule but no increased vascularity within the hypoechoic "mass-like" nodules. Findings were indeterminate for malignant change of a lipomatous lesion.

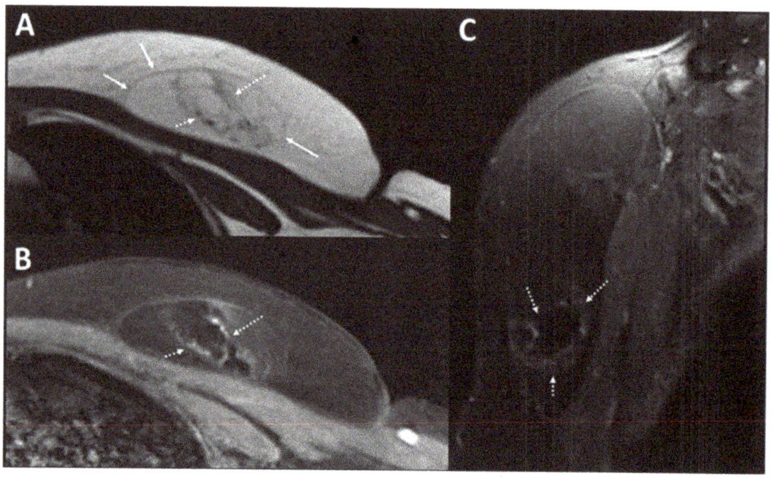


**Figure 2. MRI of left anterior chest wall.** MRI was performed for the patient in view of possible malignant change (**A**): Axial T1w image of the left anterior chest wall mass shows a fat-containing lesion in the left anterior chest wall (bold arrows) in keeping with a lipoma. Within the inferior portion, there are fat-containing areas on MRI, which are surrounded by T1w hypointense irregular bands (dotted arrows). (**B**): These irregular T1w hypointense bands show enhancement on the post contrast enhanced image. (**C**): Sagittal post contrast enhanced image shows similar findings of irregular rim enhancement around a focal fat-containing central component within inferior portions of the lipoma.

Photoacoustic (PA) tomography, a hybrid optical imaging modality, is based on the light-induced ultrasound waves providing the contrast of optical imaging combined with the high spatial resolution of ultrasound [5]. Its ability to provide the distribution of endogenous chromophores, such as blood oxygenation [6], water [7], lipid [7–10] and recently collagen [11,12], makes it attractive as a potential adjunct tool to various aspects of ultrasound. This is particularly useful in regions abundant with these chromophores, such as superficial soft tissues and the breast, or in fat-containing and fibrotic/necrotic conditions, such as fat necrosis. However, these theoretical advantages and usages of chromophore differentiations have not been demonstrated in daily clinical usage. Herein, we present interesting images that demonstrate for the first time the biochemical signatures of fat necrosis derived by PA and its agreement with histopathology (Figure 3). A detailed description of PA imaging protocol and image reconstruction is included in Appendix A. This work showcases the potential of PA as an adjunct for US to improve the diagnostic confidence for fat necrosis.

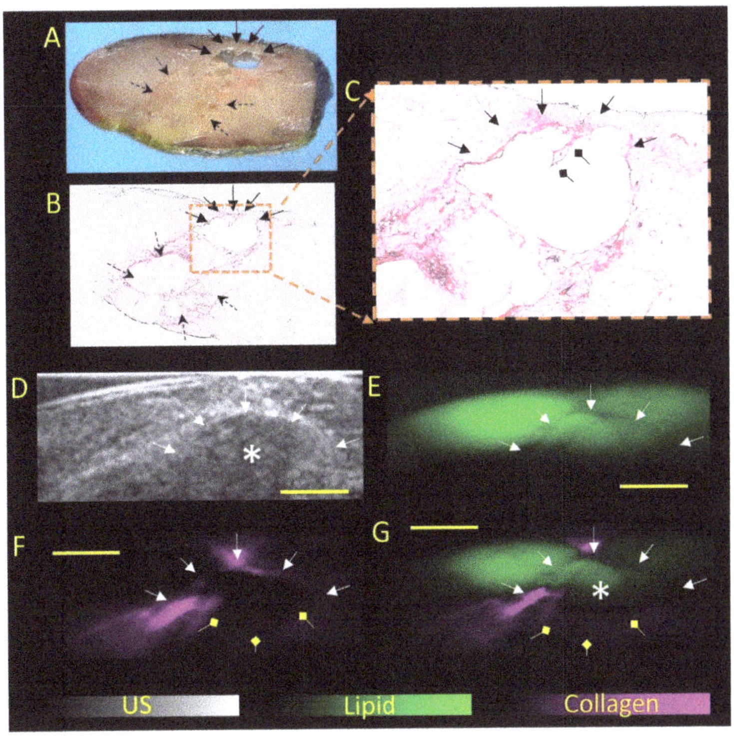


**Figure 3.** PA imaging of the excised tissue. Patient underwent uneventful excision of the left chest wall mass and was imaged with PA. Methods are as described in Appendix A. (**A**): Gross pathology of a representative cut section of the lipomatous tumor showing yellowish fatty appearance and foci of necrosis (dotted arrows). Focal cystic change containing oily fluid secondary to fat necrosis was present at where the "mass-like" nodules were seen on US (bold arrows). This area is rimmed by fibrosis. (**B**): H&E-stained microscopic image of the lipomatous tumor confirms prominent degenerative change, featuring areas of fat necrosis (dotted arrows), cystic change (bold arrows), and fibrosis. (**C**): shows zoomed in image of the cyst with septation (diamond arrow). This largest area of cystic change (bold arrows) corresponds to the "mass-like" nodule as seen on ultrasound (labelled as *). (**D**): The "mass-like" nodule near the posterior margin (*), surrounded by a hyperechoic rim (bold arrow), was targeted for PA imaging. Corresponding PA images showing distribution of lipid (**E**) and collagen (**F**), or their overlay (**G**) demonstrated collagenous signal corresponding to the hyperechoic halo in keeping with fibrosis. Within the "mass-like" region, the imaged portions demonstrated lipid signal, which was similar in intensity to the surrounding lipoma. No suspicious fat or collagenous-replacing masses were identified. No blood signals were obtained from this ex vivo study as there was no active ongoing blood flow after lesion excision. Scale bar 5 mm.

In this case, the authors have successfully demonstrated biochemical features of fat necrosis on PA as a first in-human demonstration with histopathological correlation. PA was able to identify the "mass-like" hypoechoic regions on US as fat-containing, rather than fat-replacing. On pathology, the lipid signals on PA correspond to liquefied necrotic fat from cystic degeneration, while the collagen signals on PA correspond to the fibrosis around the cavity. Hence, the biochemical capability of fat and collagen characterization could help to resolve ambiguous findings on US and improve diagnostic confidence for fat necrosis.

As it is crucial to obtain pathological correlation with US-PA images, excised tissues with no active blood signals had to be obtained. The authors believe the incorporation of blood signals in in vivo imaging would further intensify our understanding of the pathophysiology and respective imaging correlations for fat necrosis. Although the results from a single case may seem promising, more work must be done to validate these findings. In particular, more work must be done to validate findings of benign lipomas, fat necrosis, and malignant lipomatous tumors to investigate their biochemical differences. With more data, PA could potentially translate downstream into clinical imaging workflows for better characterization of superficial/breast lumps where fat-containing lesions are common. However, for widespread clinical adoption, there needs to be vast improvements in both hardware and software for PA imaging to improve its imaging depth (to at least 3–4 cm), its field of view, as well as spectral coloring.

**Author Contributions:** Conceptualization, Y.G.; methodology, Y.G., G.B., H.M.T., C.W.Q.N., E.F., R.B., M.H., S.W.T., C.W.C., M.H., T.C.P., S.A.B. and Y.T.L.; software, Y.G., G.B. and H.M.T.; writing—original draft preparation, Y.G.; writing—review and editing, Y.G., G.B. and H.M.T.; supervision, M.O. and S.T.Q. All authors have read and agreed to the published version of the manuscript.

**Funding:** This research was funded by (1) Agency of Science, Technology and Research (A*STAR), under its BMRC Central Research Fund (UIBR) 2021 and (2) NMRC clinician-scientist individual research grant new investigator grant (CS-IRG NIG).

**Institutional Review Board Statement:** The study was conducted in accordance with the Declaration of Helsinki and approved by the Institutional Review Board (or Ethics Committee) of NATIONAL UNIVERSITY HOSPITAL, SINGAPORE (2017/00805, date of approval: 12 October 2017) for studies involving humans.

**Informed Consent Statement:** Informed consent was obtained from all subjects involved in the study.

**Conflicts of Interest:** The authors declare no conflict of interest. The funders had no role in the design of the study; in the collection, analyses, or interpretation of data; in the writing of the manuscript, or in the decision to publish the results.

## Appendix A

### *Appendix A.1. Materials and Methods*

The excised lipoma was collected fresh from the operating theatre and subjected to US-PA imaging using the commercially available MSOT inVision 512 echo system (iThera Medical GmbH, Munich, Germany), fitted to a customized 2D handheld probe (specifications: 2D array of 256 detector elements (arranged along a 125° arc on a spherical surface: radius 40 mm and 5 MHz centre frequency)). The detector was placed on the inferior portions of the lipoma to image the "mass-like" nodules.

PA images were acquired at near infrared wavelengths—700, 730, 760, 800, 850, 920, 930, 970, 1000, 1050, 1064, and 1100 nm—to allow for deep-tissue imaging. Big differences in absorption of light at these wavelengths by chromophores in breast tissue, such as blood, lipid, and collagen, have aided the unmixing of these chromophores. Data were processed using ViewMSOT 3.8 software (Release 3.8, Munich, Germany) and reconstructed using the backprojection algorithm, after applying a bandpass filter with cut-off frequencies of

50 kHz and 6.5 MHz. The distribution of collagen and lipid were visualized through spectral unmixing of the reconstructed data.

## References


1. Genova, R.; Garza, R.F. *Breast Fat Necrosis*; StatPearls Publishing: Treasure Island, FL, USA, 2022.
2. Chan, L.P.; Gee, R.; Keogh, C.; Munk, P.L. Imaging features of fat necrosis. *AJR Am. J. Roentgenol.* **2003**, *181*, 955–959. [CrossRef] [PubMed]
3. Taboada, J.L.; Stephens, T.W.; Krishnamurthy, S.; Brandt, K.R.; Whitman, G.J. The many faces of fat necrosis in the breast. *AJR Am. J. Roentgenol.* **2009**, *192*, 815–825. [CrossRef] [PubMed]
4. Fernando, R.A.; Somers, S.; Edmonson, R.D.; Sidhu, P.S. Subcutaneous fat necrosis: Hypoechoic appearance on sonography. *J. Ultrasound Med.* **2003**, *22*, 1387–1390. [CrossRef] [PubMed]
5. Beard, P. Biomedical photoacoustic imaging. *Interface Focus* **2011**, *1*, 602–631. [CrossRef] [PubMed]
6. Li, M.; Tang, Y.; Yao, J. Photoacoustic tomography of blood oxygenation: A mini review. *Photoacoustics* **2018**, *10*, 65–73. [CrossRef] [PubMed]
7. Diot, G.; Metz, S.; Noske, A.; Liapis, E.; Schroeder, B.; Ovsepian, S.V.; Meier, R.; Rummeny, E.; Ntziachristos, V. Multispectral Optoacoustic Tomography (MSOT) of Human Breast Cancer. *Clin. Cancer Res.* **2017**, *23*, 6912–6922. [CrossRef] [PubMed]
8. Balasundaram, G.; Goh, Y.; Moothanchery, M.; Attia, A.; Lim, H.Q.; Burton, N.C.; Qiu, Y.; Putti, T.C.; Chan, C.W.; Hartmann, M.; et al. Optoacoustic characterization of breast conserving surgery specimens—A pilot study. *Photoacoustics* **2020**, *19*, 100164. [CrossRef]
9. Lei, P.; Hao, J.; Wang, L.; Wen, X.; Xiong, K.; Zhang, P.; Zhang, L.; Yang, S. Reliability assessment on intravascular photoacoustic imaging of lipid: Ex vivo animal and human sample validation. *J. Biophotonics* **2020**, *13*, e202000162. [CrossRef] [PubMed]
10. Zhang, J.; Yang, S.; Ji, X.; Zhou, Q.; Xing, D. Characterization of lipid-rich aortic plaques by intravascular photoacoustic tomography: Ex vivo and in vivo validation in a rabbit atherosclerosis model with histologic correlation. *J. Am. Coll. Cardiol.* **2014**, *64*, 385–390. [CrossRef] [PubMed]
11. Goh, Y.; Balasundaram, G.; Tan, H.M.; Putti, T.C.; Tang, S.W.; Ng, C.W.Q.; Buhari, S.A.; Fang, E.; Moothanchery, M.; Bi, R.; et al. Biochemical "decoding" of breast ultrasound images with optoacoustic tomography fusion: First-in-human display of lipid and collagen signals on breast ultrasound. *Photoacoustics* **2022**, *27*, 100377. [CrossRef]
12. Regensburger, A.P.; Fonteyne, L.M.; Jüngert, J.; Wagner, A.L.; Gerhalter, T.; Nagel, A.M.; Heiss, R.; Flenkenthaler, F.; Qurashi, M.; Neurath, M.F.; et al. Detection of collagens by multispectral optoacoustic tomography as an imaging biomarker for Duchenne muscular dystrophy. *Nat. Med.* **2019**, *25*, 1905–1915. [CrossRef] [PubMed]

diagnostics

Article

# Combination of Optical Biopsy with Patient Data for Improvement of Skin Tumor Identification

Yulia Khristoforova [1,*], Ivan Bratchenko [1], Lyudmila Bratchenko [1], Alexander Moryatov [2], Sergey Kozlov [2], Oleg Kaganov [2] and Valery Zakharov [1]

1 Laser and Biotechnical Systems Department, Samara National Research University, 34 Moskovskoe Shosse, 443086 Samara, Russia
2 Department of Oncology, Samara State Medical University, 89 Chapaevskaya Str., 443099 Samara, Russia
* Correspondence: khristoforovajulia@gmail.com

**Abstract:** In this study, patient data were combined with Raman and autofluorescence spectral parameters for more accurate identification of skin tumors. The spectral and patient data of skin tumors were classified by projection on latent structures and discriminant analysis. The importance of patient risk factors was determined using statistical improvement of ROC AUCs when spectral parameters were combined with risk factors. Gender, age and tumor localization were found significant for classification of malignant versus benign neoplasms, resulting in improvement of ROC AUCs from 0.610 to 0.818 ($p < 0.05$). To distinguish melanoma versus pigmented skin tumors, the same factors significantly improved ROC AUCs from 0.709 to 0.810 ($p < 0.05$) when analyzed together according to the spectral data, but insignificantly ($p > 0.05$) when analyzed individually. For classification of melanoma versus seborrheic keratosis, no statistical improvement of ROC AUC was observed when the patient data were added to the spectral data. In all three classification models, additional risk factors such as occupational hazards, family history, sun exposure, size, and personal history did not statistically improve the ROC AUCs. In summary, combined analysis of spectral and patient data can be significant for certain diagnostic tasks: patient data demonstrated the distribution of skin tumor incidence in different demographic groups, whereas tumors within each group were distinguished using the spectral differences.

**Keywords:** Raman spectroscopy; cancer risk factors; skin cancer; PLS analysis; statistical significance

**Citation:** Khristoforova, Y.; Bratchenko, I.; Bratchenko, L.; Moryatov, A.; Kozlov, S.; Kaganov, O.; Zakharov, V. Combination of Optical Biopsy with Patient Data for Improvement of Skin Tumor Identification. *Diagnostics* **2022**, *12*, 2503. https://doi.org/10.3390/diagnostics12102503

Academic Editor: Viktor Dremin

Received: 7 September 2022
Accepted: 12 October 2022
Published: 15 October 2022

**Publisher's Note:** MDPI stays neutral with regard to jurisdictional claims in published maps and institutional affiliations.

**Copyright:** © 2022 by the authors. Licensee MDPI, Basel, Switzerland. This article is an open access article distributed under the terms and conditions of the Creative Commons Attribution (CC BY) license (https://creativecommons.org/licenses/by/4.0/).

## 1. Introduction

The annually growing trend of melanoma disease is observed worldwide [1]. Research [2] estimated that 106,110 new cases of melanoma were diagnosed and about 7180 people died of this disease in the USA in 2021. The growth of melanoma can be caused by different personal [3–5], behavioral, and socioeconomic factors [6,7]. The National Cancer Institute has reported [2,8–10] that melanoma is more common in men than women and more frequent among whites in comparison with other races or ethnicities. Moreover, there is a strong relationship between melanoma cases and patient age [2]. For example, incidence rates for MM skin cancer in the UK are the highest in people aged 75 and over [11].

In terms of environmental factors, ultraviolet radiation is the most dangerous factor causing melanoma growth [11,12]. Localization can also be a potentially informative factor for more accurate skin cancer diagnosis, because some types of skin tumor often develop in the body areas that are directly exposed to UV radiation, with others appearing in covered body sites subjected to intense sunburn because of their rare exposure to regular UV radiation [13].

High risk can also be associated with family history: about 10% patients with melanoma have a family history of the disease [14–17]. The study by Hemminki et al. [18] demonstrated that melanoma is several times more common in people whose first-degree relatives have had melanoma. Moreover, researchers [19,20] reported on the relationship between patient history of skin neoplasm and the risk of developing melanoma, suggesting that the patient's history indicates risk of skin cancer growth. People working in certain professions can have a higher risk of skin cancer and some precancer conditions, due to interaction with dangerous industrial carcinogens [21].

Preliminary diagnosis of melanoma using dermoscopy [22] or other developing optical biopsy techniques [23–25] did not consider the above risk factors that may be potential prerequisites for developing skin cancer. However, incorporating patient-specific information can improve the accuracy of disease identification based on clinical studies [26–28]. Pacheco and Krohling [26] demonstrated the importance of clinical features for skin cancer detection based on clinical images and confirmed the hypothesis that patient clinical information is important for this task. However, they concluded that the clinical features they examined were not practical indicators for all types of skin lesions. Zeng et al. [27] examined skin tumors using Raman spectral data, considering various risk factors, and revealed that only patient age significantly contributed to improved diagnosis of malignant tumors. Taking into account the findings of other research teams [26–28], we aimed to test the possibility of improving skin cancer identification with our experimental data, by combining Raman and autofluorescence data as well as patient information.

In our previous work [29,30], we performed an optical biopsy using Raman and autofluorescence (AF) spectroscopy to diagnose skin cancer. Raman spectroscopy has proved to be a sensitive research instrument in clinical practice for a number of purposes [24,27,30,31]. The proposed method [29,32] was able to classify skin neoplasms with a mean accuracy higher than the accuracy of general practitioners or trainees, and with comparable or less accuracy than trained dermatologists and experts. Therefore, it remains necessary to improve the accuracy of skin cancer diagnosis performed with Raman and AF analysis.

The aim of our study was to estimate the prognostic possibility of combining individual patient factors with the results of optical biopsy for detecting skin cancer. The spectral data of 617 skin tumors that were analyzed in our previous work [29] were combined with data on risk factors, for joint analysis. We demonstrated the results of the proposed approach by combining Raman spectroscopy and AF with individual patient factors such as environmental risk, history, and personal risk factors for classification of malignant skin tumors, melanoma, and other skin neoplasms.

## 2. Materials and Methods

### 2.1. Experimental Setup

The detailed description of the experimental setup for simultaneous Raman and AF signal registration was presented in our previous studies [29,32]. The scattering spectral response from skin tissue in the near-infrared region was stimulated using a thermally stabilized diode laser module (LuxxMaster, LML-785.0RB-04, PD-LD, Ushio Inc., Tokyo, Japan) with 785 $\pm$ 0.1 nm central wavelength. The laser power density on the skin was about 0.3 W/cm$^2$ and did not cause any damage to skin or discomfort in patients. The optical Raman probe (RPB785, InPhotonics Inc., Norwood, MA, USA) contained supplying and collecting branches. Laser excitation at 785 nm was delivered to the skin surface by means of the excitation optic fiber (0.22 NA, 100 µm) and the supplying branch of the probe with a band-pass filter and a focusing lens. The scattered radiation was collected by the same lens and delivered to the collecting branch by the dichroic mirror and the conventional mirror. The longpass filter cut the excitation laser wavelength from the collected signal, and the Raman and fluorescence signals of skin tissue were transmitted to the spectrometer using the focusing lens and the collecting fiber (NA 0.22, 200 µm). The collected signal was decomposed into a spectrum using a portable spectrometer (QE65Pro, Ocean Optics Inc., Largo, FL, USA). The spectra were registered in the 780–1000 nm region with spectral resolution of 0.2 nm. The acquisition time was 20 s with a triple accumulation. The QE65Pro detector was cooled down to −15 °C. The silicon tip on the probe provided the 7–8 mm distance between the skin surface and the probe for all measurements.

### 2.2. Patients

The protocols of the in vivo tissue diagnostics were approved by the ethical committee of Samara State Medical University (Samara Region, Samara, Russia, protocol No 132, 29 May 2013), the clinical studies fall within The Code of Ethics of a Doctor of Russia, approved at the 4th Conference of the Russian Medical Association, and within the World Medical Association Declaration of Helsinki. The study involved 615 patients of different ages, including 178 men and 437 women, who consulted specialized oncologists in the Samara Regional Clinical Oncology Dispensary from May 2017 to December 2019. All the patients were aged $\geq 18$. Informed consent was acquired from all patients before the in vivo study.

Spectral measurements of 617 tumors were carried out for 615 patients. The spectral measurement of each skin tumor was registered from the approximate central point of the tumor area. The region of interest for spectral registration of tumors was confirmed by a medical specialist on the basis of dermatoscopic images. The skin tumors were localized at different body sites. The sizes of skin tumors varied widely, from 0.3 to 5 cm. Summary of the patients and tumors is presented in [29]. In accordance with results of histopathological analysis, the analyzed spectral cohort included 204 malignant tumors (70 malignant melanomas (MM), 122 basal cell carcinomas (BCC) and 12 squamous cell carcinomas (SCC)), as well as 413 benign tumors (26 dermatofibromas (DF), 62 papillomas (PP), 40 hemangiomas (HE), 113 seborrhoeic keratosis (SK), 170 nevi (NE) (all types), 1 cutaneous horn, and 1 benign tumor of epidermal appendage).

### 2.3. Risk Factors for Skin Cancer Growth

Cancer develops when human cells are damaged due to various factors and the number of damaged cells starts to grow uncontrollably. In this work, we analyzed several risk factors that can potentially provoke skin cancer growth.

At the initial appointment, the oncologist collected the patient history and potential risk factors for skin cancer growth: gender (G), age (A), tumor localization (L), family history (FH), personal history (PH), sun exposure (SE), size (S), and occupational hazards (OH). All the collected demographic indicators were defined by the patient survey. However, for different reasons, not every patient provided the full set of collected risk factors. Only gender, age and localization factors were received for all the 617 skin neoplasm spectra. Therefore, we considered two spectra datasets:

(I) Spectral data of all 617 skin neoplasms with only three indicators: (G), (A), (L);
(II) Spectral data of only 481 out of the 617 skin neoplasms with all eight indicators: (G), (A), (L), (FH), (PH), (SE), (S), (OH).

All the risk factors were digitized:

G: 1—male; 2—female;
A: 1—under 29, 2—30 to 39, 3—from 40 to 49, 4—from 50 to 59, 5—from 60 to 69, 6—over 70;
L: 1—head and neck, 2—trunk, 3—upper limb, 4—lower limb;
FH: 0—no malignant diseases in close relatives; 1—close relatives with malignant diseases, 2—close relatives with skin cancer disease;
PH: 0—the patient had no serious disease; 1—the patient had a different disease; 2—the patient had a malignant disease;
SE: 0—the patient avoids suntan; 1—the patient gets suntan without sunburn; 2—the patient often has sunburn;
S: 1—from 0 to 5 mm; 2—from 6 to 20 mm; 3—21 mm;
OH: 0—no occupational hazards; 1—occupational hazards due to skin contact with chemicals (e.g., work with petroleum products, on chemical plants, etc.).

The digitization of the patient factors was performed by specialized oncologists at the Samara Regional Clinical Oncology Dispensary.

### 2.4. Preprocessing and Statistical Analysis of Spectra

The spectra were recorded in the 780–1000 nm region, but only the 803–914 nm spectral region corresponding to the 300–1800 $cm^{-1}$ wavenumber region in terms of Raman spectroscopy was analyzed. Firstly, the raw spectra in the region of interest (803–914 nm) were preprocessed by the following process: smoothing by the Savitsky–Golay filter, normalization by the standard normal variate method (SNV), and centering.

In accordance with the data described in Section 2.2, we considered six classification models with different sets of risk factors:

I.1 Malignant (n = 204) vs. benign (n = 413) neoplasms with 3 risk factors;
II.1 Malignant (n = 157) vs. benign (n = 324) neoplasms with 8 risk factors;
I.2 MM (n = 70) vs. benign pigmented (Ne and SK, n=283) neoplasms with 3 risk factors;
II.2 MM (n = 49) vs. benign pigmented (Ne and SK, n = 221) neoplasms with 8 risk factors;
I.3 MM (n = 70) vs. SK (n = 113) with 3 risk factors;
II.3 MM (n = 49) vs. SK (n = 90) with 8 risk factors.

Each spectrum included the Raman and AF signals in the region of interest (of 803–914 nm) and, therefore, represented a discrete set of intensity values at the 515 wavelengths (in accordance with the spectral resolution of the spectrometer). For the subsequent regression analysis, the 515 spectral parameters respectively representing each tumor after preprocessing were combined with the corresponding risk factor parameters. Therefore, in classification models (I.1), (I.2), and (I.3) each tumor was represented as 518 predictors (515 spectral parameters and three risk factor parameters) for PLS analysis, and in classification models (II.1), (II.2), and (II.3) as 523 predictors (515 spectral parameters and eight risk factor parameters), respectively.

The experimental data were processed using partial least square discriminant analysis (PLS-DA) [33]. The PLS-DA method was applied to build a regression model between the analyzed tumor predictors and tumor types. Stability of the PLS-DA classification was checked by means of 10-fold cross-validation. The number of latent variables (LVs) for the PLS-DA models was chosen according to the minimum of the RMSE in the 10-fold cross-validation. To estimate the importance of all tumor predictors in the model, variable importance in projection (VIP) analysis was performed [34]. The VIP scores highlighted the informative predictors of tumors in the regression model that were more important for classifying different tumor types. Higher relative intensity of VIP score indicated that the predicted variable was more significant. To determine the differentiation accuracy of the tumor analysis, the PLS predictors were calculated as numeric values of tumor diagnosis in the built regression model.

The results of the skin tumor differentiation were visualized using a bee-swarm diagram and the receiver operating characteristic (ROC) curves plotted using R studio software [35]. The ROC analysis shows the diagnostic performances of the regression model. For quantitative analysis, the area under the curve (AUC) was calculated. The significance of the AUCs and the comparisons between different AUCs were tested in a standard manner [36].

## 3. Results

### 3.1. Malignant vs. Benign Neoplasms

(I.1) To discriminate the malignant (n = 204) vs. benign (n = 413) neoplasms from set (I), the 0.600 (0.567–0.652) ROC AUC was obtained using only the spectral data (RS and AF data). The complementation of spectral dataset (I) with three risk factors made it possible to improve the ROC AUCs to 0.818 (0.778–0.841). Moreover, adding each patient factor separately to the spectral data significantly increased the ROC AUC (see Table 1). The distribution of VIP scores as a weighted sum of loadings is shown in Figure 1, highlighting all spectral features for all loadings obtained in this PLS classification model. For this model, the VIP scores were utilized to classify malignant versus benign tumors by determination of informative predictors (gender (G), age (A), location (L), and 515 spectral parameters) in regression specification. The VIP scores presented in Figure 1 demonstrate that age (A) is the most informative risk factor, which was proved by the most significant improvement of ROC AUC (0.804, $p = 9 \times 10^{-9}$) when only age was incorporated into the spectral data, compared with the other factors.

**Table 1.** Results of regression models.

| Model | ROC AUC |
|---|---|
| **I.1 Malignant (n = 204) vs. Benign (n = 413), cohort with 3 risk factors** | |
| only spectral data (803–914 nm) | 0.600 (0.567–0.652) |
| spectral data with gender | 0.691 (0.647–0.736), $p = 0.008$ |
| spectral data with age | 0.804 (0.767–0.840), $p = 9 \times 10^{-9}$ |
| spectral data with localization | 0.759 (0.718–0.800), $p = 3 \times 10^{-6}$ |
| spectral data with all risk factors | 0.818 (0.778–0.841), $p = 2 \times 10^{-11}$ |
| **II.1 Malignant (n = 157) vs. Benign (n = 324), cohort with 8 risk factors** | |
| only spectral data (803–914 nm) | 0.610 (0.556–0.663) |
| spectral data with gender | 0.707 (0.658–0.756), $p = 0.006$ |
| spectral data with age | 0.718 (0.671–0.766), $p = 0.002$ |
| spectral data with localization | 0.680 (0.628–0.732), $p = 0.035$ |
| spectral data with family history | 0.625 (0.570–0.677), $p = 0.35$ |
| spectral data with personal history | 0.609 (0.556–0.663), without improvement |
| spectral data with sun exposure | 0.609 (0.555–0.663), without improvement |
| spectral data with size | 0.689 (0.639–0.738), $p = 0.02$ |
| spectral data with occupational hazards | 0.616 (0.563–0.669), $p = 0.43$ |
| spectral data with all risk factors | 0.789 (0.746–0.832), $p = 5 \times 10^{-7}$ |
| **I.2 MM (n = 70) vs. Ne + SK (n = 283), cohort with 3 risk factors, n = 353** | |
| only spectral data (803–914 nm) | 0.690 (0.630–0.761) |
| spectral data with gender | 0.751 (0.685–0.818), $p = 0.2$ |
| spectral data with age | 0.771 (0.706–0.837), $p = 0.1$ |
| spectral data with localization | 0.772 (0.709–0.835), $p = 0.1$ |
| spectral data with all risk factors | 0.825 (0.766–0.884), $p = 0.02$ |
| **II.2 MM (n = 49) vs. Ne + SK (n = 221) (cohort with 8 risk factors, n = 270)** | |
| only spectral data (803–914 nm) | 0.789 (0.718–0.861) |
| spectral data with gender | 0.801 (0.729–0.873), $p = 0.4$ |
| spectral data with age | 0.808 (0.734–0.881), $p = 0.37$ |
| spectral data with localization | 0.804 (0.737–0.871), $p = 0.4$ |
| spectral data with family history | 0.796 (0.726–0.866), $p = 0.45$ |
| spectral data with personal history | 0.744 (0.668–0.819), without improvement |
| spectral data with sun exposure | 0.798 (0.725–0.870), $p = 0.44$ |
| spectral data with size | 0.806 (0.736–0.876), $p = 0.38$ |
| spectral data with occupational hazards | 0.788 (0.714–0.861), without improvement |
| spectral data with all risk factors | 0.849 (0.785–0.914), $p = 0.14$ |
| **I.3 MM (n = 70) vs. SK (n = 113) (cohort with 3 risk factors, n = 183)** | |
| only spectral data (803–914 nm) | 0.791 (0.728–0.859) |
| spectral data with gender | 0.791 (0.722–0.859), without improvement |
| spectral data with age | 0.791 (0.723–0.859), without improvement |
| spectral data with localization | 0.841 (0.783–0.900), $p = 0.15$ |
| spectral data with all risk factors | 0.844 (0.786–0.902), $p = 0.15$ |
| **II.3 MM (n = 49) vs. SK (n = 90) (cohort with 8 risk factors, n = 139)** | |
| only spectral data (803–914 nm) | 0.814 (0.740–0.888) |
| spectral data with gender | 0.815 (0.741–0.889), $p = 0.49$ |
| spectral data with age | 0.815 (0.740–0.889), $p = 0.49$ |
| spectral data with localization | 0.851 (0.784–0.918), $p = 0.25$ |
| spectral data with family history | 0.816 (0.743–0.889), $p = 0.48$ |
| spectral data with personal history | 0.815 (0.742–0.889), $p = 0.49$ |
| spectral data with sun exposure | 0.815 (0.741–0.889), $p = 0.49$ |
| spectral data with size | 0.860 (0.795–0.925), $p = 0.19$ |
| spectral data with occupational hazards | 0.815 (0.740–0.889), $p = 0.49$ |
| spectral data with all risk factors | 0.820 (0.748–0.892), $p = 0.46$ |

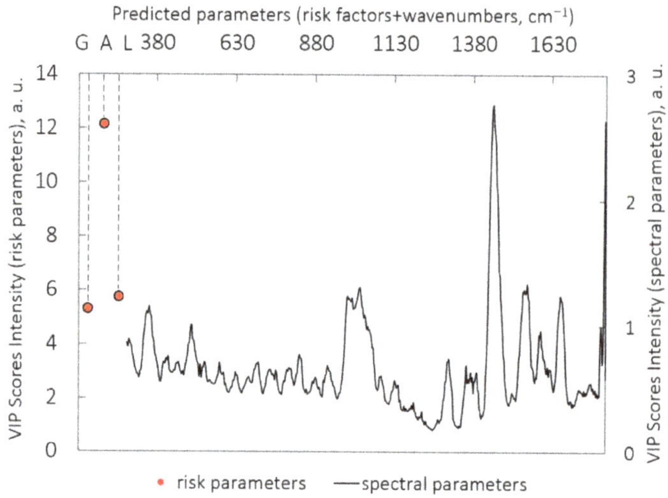


**Figure 1.** VIP scores for PLS-DA model: classification of malignant (n = 204) vs. benign (n = 413) neoplasms (I.1); importance values of risk factors and spectral factors are plotted along different horizontal axes because of a wide value scatter. (G) gender; (A) age; (L) localization.

The ROC AUCs and the bee-swarm diagram for this classification are presented in Table 1 and Figure 2a–c.

(II.1) For set (II), the classification of malignant (n = 157) vs. benign (n = 324) neoplasms using the PLS analysis was performed with the 0.610 (0.556–0.663) ROC AUC on the basis of only the spectral data, and with the 0.789 (0.746–0.832) ROC AUC when supplying the spectral data with eight risk factors. For this set, age was also the most important risk factor. The ROC AUCs and bee-swarm diagram are presented in Table 1 and Figure 2d–f.

Table 1 presents the ROC AUCs of the models built using all risk factors separately. Improvement of the ROC AUC by incorporating the spectral data with all risk factors to identify malignant skin cancer was statistically significant ($p < 0.05$) in models I.1 and II.2.

### 3.2. MM vs. Benign Pigmented Neoplasms (Ne and SK)

(I.2) In this classification task, regression analysis of the cases from dataset (I) using only the spectral data was performed with 0.690 (0.630–0.761) ROC AUC. For this task, the combined analysis of the spectral data and the three risk factors significantly improved the diagnostic performance to 0.825 (0.766–0.884) ROC AUC. The contribution of all three risk factors in this model was significant for MM identification ($p < 0.05$), whereas separately adding age, gender, or localization did not result in significant improvement of the ROC AUC. Figure 2g–i and Table 1 show the results from cohort (I) for this classification task.

(II.2) In the same classification task for cohort (II), MMs (n = 49) were differentiated from benign pigmented neoplasms (n = 221) with 0.789 (0.718–0.861) ROC AUC using only the Raman and AF spectral data, and 0.849 (0.785–0.914) ROC AUC when combining the spectral and risk factor variables. However, in this case, there were no significant differences ($p = 0.14$) between the ROC AUCs obtained for the models with the eight risk factors and those without. Figure 2 presents only the statistically significant results and therefore does not include a diagram for this model. Table 1 indicates the ROC AUCs for this classification task.

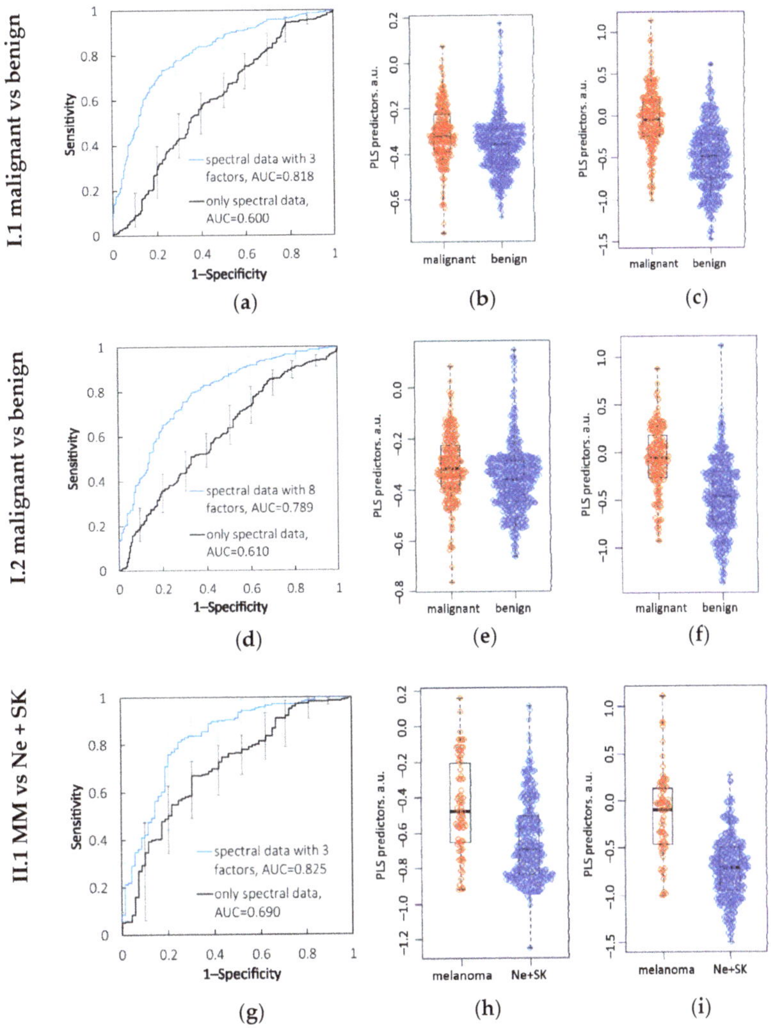


**Figure 2.** Results of diagnostic models with statistical significance ($p < 0.05$). I.1: Malignant vs. benign neoplasms: (**a**) ROC AUCs, bee-swarm diagrams of PLS predictors for tumor classification based on (**b**) spectral parameters and (**c**) combination of spectral parameters and three patient factors. II.1: Malignant vs. benign neoplasms: (**d**) ROC AUCs, bee-swarm diagrams of PLS predictors for tumor classification based on (**e**) spectral parameters and (**f**) combination of spectral parameters and eight patient factors. I.2: MM vs. benign pigmented neoplasms: (**g**) ROC AUCs, bee-swarm diagrams of PLS predictors for tumor classification based on (**h**) spectral parameters and (**i**) combination of spectral parameters and three patient factors.

### 3.3. *MM vs. SK*

(I.3) When classifying the MMs (n = 70) vs. SKs (n = 113) from cohort (I) using the spectral data, 0.791 (0.722–0.859) ROC AUC was obtained. When the three risk factors were combined with the Raman and AF spectral data, the ROC AUC increased to 0.844 (0.786–0.902) but with no statistical significance.

(II.3) When analyzing dataset (II) with the same classification task, the discrimination model showed 0.820 (0.748–0.892) ROC AUC obtained by combining spectral data and the eight risk factors, while the analysis of only the spectral data was performed with 0.814 (0.740–0.888) ROC AUC.

The AUCs of this classification task were insignificantly improved ($p > 0.05$) in models I.3 and II.3. The diagnostic performance results for this classification task are presented in Table 1.

## 4. Discussion

The classification results for different types of skin neoplasms based on Raman and AF spectral data were demonstrated in our previous research [29,32]. In this current work, we combined individual patient risk factors and spectral data to obtain a more precise skin cancer diagnosis, in particular of malignant tumor and MM. It should be noted that true statistics of skin cancer incidence might differ from the data we obtained, due to including in this study only those patients who were aware of skin tumors, attentive to their health, and had access to resources for tumor detection.

Considering the significance of the analyzed patient factors, our classification models have demonstrated that although gender was a significant factor for classifying malignant versus benign skin tumors in both model I.1 and model II.1, it was not significant for diagnosis of MM. The analyzed cohort was heterogeneous in the numbers of men and women: women outnumbered men 2–3 times in the general cohort and in the analyzed classes. However, the proportion of men with malignant tumors among all men involved in this study was 0.43, whereas the relative proportion of women was 0.28. At the same time, the relative incidence rates of MM were 0.13 among men and 0.11 among women (Figure S1a). The statistics for different ethnicities and races vary. According to statistics from Australia and the USA [10,13], the incidence of MM is higher among men than women. In Russia in 2020, the standardized incidence rates (number per 100,000 population) of skin cancer (without MM) and MM among men were 21.48 and 4.08 respectively, whereas among women the figures were 20.62 (skin cancer without MM) and 4.32 (MM) [37]. Data for our cohort was collected from May 2017 to December 2019, and revealed that in our study the relative number of malignant cases was higher among men, while the number of MM cases was the same among men and women.

Localization was also a significant factor in models I.1 and II.1 for malignant versus benign classification. We suppose that the significance of this factor can be explained by the most common localization group for different tumor types (Figure S1b). In model I.1, most of the malignant tumors, namely 86 out of 204 cases (about 42%), were located on the head and neck, while 209 out of 413 benign tumors (about 51%) were situated on the trunk. Despite the fact that more MM cases occurred on the trunk (about 51%), the large number of malignant tumors on the head and the neck was due to the contribution of the BCC and SCC cases. However, when classifying MM and benign pigmented tumors or SK, this factor was found to be insignificant, because most cases within each class occurred on the trunk: 51% among melanoma cases, 51% among benign pigmented cases, and 47% among cases of seborrheic keratosis.

The sun exposure factor was insignificant in all models (II.1, II.2, and II.3), but exposure to sun radiation is partly responsible for localization. BCC and SCC are more likely to occur on body areas that are most exposed to solar radiation, i.e., on the head. According to our data presented in Figure S1, 61% of BCC and 71% of SCC in this study were localized on the head and the neck [29]. On the other hand, most MM (51%) and other melanocytic tumors, such as pigmented nevus, occurred on the trunk and legs: these body areas may be subjected to intense sunburn because of less frequent exposure to regular UV radiation. Other research [38] reports that trunk melanomas are more strongly associated with pigmented nevus counts. Thus, exposure to sun radiation is an equally important growth factor for all melanocytic tumors, confirmed by a similar distribution of various melanocytic tumors (e.g., most cases of MM and benign pigmented nevus were recorded on the trunk). Therefore, these tumors were not distinguished within each localization.

Our models suggest that age is a significant factor when classifyin malignant and benign tumors, because the patient distribution in each age group was different. Most cases of malignant tumors (about 44%) were recorded in patients over 70, while the groups of patients with benign tumors aged from 30 to 39, from 40 to 49, from 50 to 59, and from 60 to 69 were equal in size. For classification of MM versus only SK, age was completely insignificant and did not improve the ROC AUC when added to the spectral data, because the distribution of patients in age groups for MM and SK was fairly similar (Figure S1c). The maximum frequency of MM and benign pigmented tumors (Ne and SK) was recorded in the age group from 60 to 69. The larger number of pigmented benign tumors in the age group from 60 to 69 was due to more SK cases, which did not allow significant separation of these classes by age. However, there were differences in the numbers of patients in other age groups: among those with benign pigmented tumors, more than 30% were under 40 because of a larger number of young patients with pigmented nevus. This resulted in the fact that adding the age factor to the spectral data improved the ROC AUC to 0.808 (0.734–0.881) for the classification of MM versus benign pigmented tumors, but these differences were not sufficient for statistical significance.

Finally, it should be noted that in all three classification tasks (models II.1, II.2, and II.3), OH, FH, and PH were not able to improve the ROC AUCs (see Table 1). We lack precise data on the behavioral and genetic factors obtained in the survey because the oncologists collected this information by questioning the patients, which can lead to inaccuracy and uncertainty. To enhance the importance of these factors, they could be defined in a more precise way, for example, in terms of controlling exposure to sunlight or to chemicals that can be dangerous in case of contact with skin. Thereby, the significance of such patient data as gender, age, tumor localization, and size can become more reliable for tumor detection.

Diagnostic performance combining patients' demographic data with optical data has been evaluated in several works [26–28]. Zhao et al. [27] investigated whether incorporating such patient demographics as gender, skin type, localization, and age into Raman spectral analysis can improve performance in malignant skin cancer diagnosis. Using PLS analysis, the authors reported that the ROC AUC improved significantly from 0.913 (0.892–0.933) to 0.934 (0.917–0.952) ($p < 0.05$) after combining only the Raman data with all the demographics, to differentiate malignant and benign skin lesions. In comparison with the study by Zeng et al., in our work we analyzed a larger set of risk factors for cancer growth, including not only demographics but patient lifestyle and behavioral factors as well. We similarly found that combining all the risk factors with the spectral data achieved better performance in discriminating malignant and benign tumors, increasing the ROC AUC from 0.600 to 0.818 with three factors and from 0.610 to 0.789 with eight factors. We increased the ROC AUCs by 30–36% taking into account the patient risk factors, while in the study by Zeng et al. [27] the improvement was only 2%. Probably, this greater improvement of malignant skin cancer identification was a result of a different signal-to-noise ratio in the spectral data. Our spectral data were recorded with a lower signal-to-noise ratio [39] that resulted in low accuracy of detecting malignant neoplasms by only the Raman and AF spectra (0.600 and 0.610 ROC AUC in models I.1 and II.1, respectively), and a significant improvement when the patient factors were added. In the previous work [27], the authors used a highly sensitive spectroscopic system that allowed them to obtain a high ROC AUC 0.913 using only the Raman spectral data.

Kharazmi et al. [28] proposed a non-invasive fast BCC detection tool that incorporates dermoscopic lesion features and clinical patient information including lesion localization, size, and elevation, as well as patient age and gender. The integrated analysis of the patient profile and dermoscopic features using data-driven feature learning allowed them to increase the ROC AUC for BCC detection from 0.847 to 0.911, in comparison with only the dermoscopic features. According to our statistics [29], BCC cases occur within specific demographic conditions, for example, 61% of BCCs were located on the head and neck, and 90% of BCCs were recorded from patients over 60. Thus, it might be assumed that we would be able to determine the significance of patient factors for BCC detection. However, we analyzed BCC and MM as malignant tumors and did not estimate the importance of patient information for identifying BCC only. Considering that our statistical results about BCC are in good agreement with the statistics reported by Kharazmi et al., this may suggest the significance of patient factors only for several types of skin lesions.

It is also interesting to compare the results of the proposed methodology and the results of dermoscopic image analysis performed by dermatologists, which represents the current standard for clinical diagnosis of skin lesions. According to research [40], the accuracy of melanoma vs. non-melanoma skin lesion classification was 79.9% for novice dermatologists, 83.3% for qualified dermatologists, and 86.9% for experts. The mean diagnostics performance of 21 board-certified dermatologists using dermoscopic images to classify 71 malignant vs. 40 benign lesions was nearly 71% sensitivity and 81% specificity [41]. Thus, the proposed methodology can classify skin neoplasms with a mean accuracy higher than GPs and trainees, but with slightly less accuracy than trained dermatologists and experts.

To sum up, our results show that information on patient risk factors and Raman and AF spectral data can complement each other to provide more accurate skin cancer identification. For each skin tumor type, we observed a specific distribution trend by gender, age group, and localization, in good agreement with worldwide statistics on skin tumor incidence. Patients' age and tumor localization are able to discriminate tumors in different groups, but these factors become insignificant when analyzing different skin tumors within individual groups. For example, similar numbers of malignant, benign, pigmented tumors, SK, and MM were recorded on the trunk or in patients aged from 50 to 59 or from 60 to 69. So, within a separate demographic group, accuracy results for different tumor type diagnosis can differ when only the Raman and AF spectral data are used. For this reason, to differentiate malignant versus benign skin tumors we improved the ROC AUCs by adding risk factors to the model. To differentiate MM versus pigmented skin tumors or SK, similar demographic trends did not allow us to increase the performance accuracy of skin tumor identification. To improve diagnostic performance, the proposed methodology may be added to the estimation of neoplasm morphology performed during dermoscopy analysis. Deep learning-based applications using computer visualization have shown promising results in detecting melanoma based on the analysis of dermoscopic images [40–42]. However, additional studies are required to estimate the capability of joint dermoscopy analysis and low-cost Raman systems.

## 5. Conclusions

We tested the possibility of improving skin cancer detection by combining spectral analysis with analysis of individual patient characteristics and factors for skin cancer growth. We analyzed two cohorts of patients with skin tumors: (I) the cohort with 617 spectra of different tumors and three patient factors for each case, and (II) the cohort with 481 spectra of different tumors and eight risk factors. For each cohort, three classification tasks were considered: malignant versus benign tumors, MM versus benign pigmented tumors, and MM versus SK.

The significance of risk factors for type of cancer growth was estimated when all factors were combined with the spectral data, and when each factor was added separately to the Raman and AF spectral data. Statistical improvement was achieved for the classification of 204 malignant tumors and 413 benign tumors, from 0.610 to 0.818 ROC AUC, $p = 2 \times 10^{-11}$, when spectral data in the 300–1800 cm$^{-1}$ range were combined with three individual patient factors for skin cancer growth. Moreover, classification of 157 malignant tumors and 324 benign tumors using the spectral data and eight risk factors was statistically improved from 0.610 to 0.789, $p = 5 \times 10^{-7}$. Finally, 70 MMs and 283 benign pigmented skin neoplasms were differentiated with a statistical improvement from 0.709 to 0.825, $p = 0.02$ when combining the spectral data and the three risk factors. Improvements of ROC AUC for discriminating MM (n = 49) and pigmented benign tumors (n = 172) with eight factors, MM (n = 70) and SK (n = 113) with three factors, and the MM (n = 49) and SK (n = 90) with eight factors were all statistically insignificant.

Our results show that among all risk factors, patient demographics including gender, age, and tumor localization were statistically significant for detecting skin tumor type, due to their univocal definition. In contrast, the data for behavioral factors were collected by staff directly from patients and might therefore lack accuracy. For certain classification tasks, it was found that the combination of spectral data and patient risk factors was significant. Particular overall trends for each skin tumor type were observed for patient age, gender, and tumor localization. However, these demographic features did not allow us to discriminate different tumor types, especially pigmented tumors, within an individual demographic group. Therefore, distinguishing skin tumors in groups with similar demographics was possible using the Raman and AF spectral data only. However, these findings need to be verified in further experimental cohort studies.

**Supplementary Materials:** The following supporting information can be downloaded at: https://www.mdpi.com/article/10.3390/diagnostics12102503/s1, Figure S1: Patients' statistics: (**a**) distribution by gender; (**b**) distribution by localization; (**c**) distribution by age groups.

**Author Contributions:** Conceptualization, I.B., V.Z.; Methodology, Y.K., I.B., L.B.; Validation, I.B., V.Z.; Investigation, Y.K., L.B., A.M.; Resources, S.K., V.Z.; Data Curation, I.B., A.M.; Writing—Original Draft Preparation, Y.K.; Writing—Review & Editing, I.B., V.Z.; Supervision, V.Z., S.K., O.K.; Project Administration, V.Z., O.K.; Funding Acquisition, I.B. All authors have read and agreed to the published version of the manuscript.

**Funding:** This work was supported by the Russian Science Foundation under Grant No. 21-75-10097, https://rscf.ru/project/21-75-10097/.

**Institutional Review Board Statement:** The protocols of the in vivo tissue diagnostics were approved by the ethical committee of Samara State Medical University (Samara Region, Samara, Russia, protocol No 132, 29 May 2013); clinical studies fall within The Code of Ethics of a Doctor of Russia, approved at the 4th conference of the Russian Medical Association, and within the World Medical Association Declaration of Helsinki.

**Informed Consent Statement:** Informed consent was obtained from all subjects involved in the study.

**Data Availability Statement:** The data presented in this study are available on request from the corresponding author.

**Conflicts of Interest:** The authors declare no competing interests.

## References

1. Sung, H.; Ferlay, J.; Siegel, R.L.; Laversanne, M.; Soerjomataram, I.; Jemal, A.; Bray, F. Global Cancer Statistics 2020: GLOBOCAN Estimates of Incidence and Mortality Worldwide for 36 Cancers in 185 Countries. *CA Cancer J. Clin.* **2021**, *71*, 209–249. [CrossRef] [PubMed]
2. National Cancer Institute. Cancer Stat Facts: Melanoma of the Skin. Available online: https://seer.cancer.gov/statfacts/html/melan.html (accessed on 28 October 2021).
3. American Cancer Society. Risk Factors for Melanoma Skin Cancer. Available online: https://www.cancer.org/cancer/melanoma-skin-cancer/causes-risks-prevention/risk-factors.html (accessed on 28 October 2021).
4. Leonardi, G.C.; Falzone, L.; Salemi, R.; Zanghì, A.; Spandidos, D.A.; Mccubrey, J.A.; Candido, S.; Libra, M. Cutaneous melanoma: From pathogenesis to therapy. *Int. J. Oncol.* **2018**, *52*, 1071–1080. [CrossRef] [PubMed]
5. Ossio, R.; Roldán-Marín, R.; Martínez-Said, H.; Adams, D.J.; Robles-Espinoza, C.D. Melanoma: A global perspective. *Nat. Rev. Cancer* **2017**, *17*, 393–394. [CrossRef] [PubMed]
6. Risk Factors and Cancer. Global Cancer Observatory Global Cancer Observatory. Available online: https://gco.iarc.fr/databases.php (accessed on 28 October 2021).
7. Johnson-Obaseki, S.E.; Labajian, V.; Corsten, M.J.; McDonald, J.T. Incidence of cutaneous malignant melanoma by socioeconomic status in Canada: 1992–2006. *J. Otolaryngol. Head Neck Surg.* **2015**, *44*, 53. [CrossRef] [PubMed]
8. Neale, R.E.; Davis, M.; Pandeya, N.; Whiteman, D.C.; Green, A.C. Basal cell carcinoma on the trunk is associated with excessive sun exposure. *J. Am. Acad. Dermatol.* **2007**, *56*, 380–386. [CrossRef]
9. Gupta, A.K.; Bharadwaj, M.; Mehrotra, R. Skin cancer concerns in people of color: Risk factors and prevention. *Asian Pac. J. Cancer Prev. APJCP* **2016**, *17*, 5257.

10. United States Cancer Statistics: Data Visualizations. Leading Cancers by Age, Sex, Race and Ethnicity. Available online: https://gis.cdc.gov/Cancer/USCS/#/Demographics/ (accessed on 28 October 2021).
11. Whiteman, D.C.; Stickley, M.; Watt, P.; Hughes, M.C.; Davis, M.B.; Green, A.C. Anatomic site, sun exposure, and risk of cutaneous melanoma. *J. Clin. Oncol.* **2006**, *24*, 3172–3177. [CrossRef]
12. Cancer Research UK. Causes of Cancer and Reducing Your Risk. Available online: https://www.cancerresearchuk.org (accessed on 28 October 2021).
13. Raimondi, S.; Suppa, M.; Gandini, S. Melanoma epidemiology and sun exposure. *Acta Derm. Venereol.* **2020**, *100*, 250–258. [CrossRef]
14. Diseati, L.; Scheinfeldt, L.B.; Kasper, R.S.; Zhaoyang, R.; Gharani, N.; Schmidlen, T.J.; Gordon, E.S.; Sessions, C.K.; Delaney, S.K.; Jarvis, J.P.; et al. Common genetic risk for melanoma encourages preventive behavior change. *J. Pers. Med.* **2015**, *5*, 36–49. [CrossRef]
15. Gu, F.; Chen, T.H.; Pfeiffer, R.M.; Fargnoli, M.C.; Calista, D.; Ghiorzo, P.; Landi, M.T. Combining common genetic variants and non-genetic risk factors to predict risk of cutaneous melanoma. *Hum. Mol. Genet.* **2018**, *27*, 4145–4156. [CrossRef]
16. Meyle, K.D.; Guldberg, P. Genetic risk factors for melanoma. *Hum. Genet.* **2009**, *126*, 499–510. [CrossRef] [PubMed]
17. Gordon, R. Skin cancer: An overview of epidemiology and risk factors. *Semin. Oncol. Nurs.* **2013**, *29*, 160–169. [CrossRef] [PubMed]
18. Hemminki, K.; Zhang, H.; Czene, K. Familial and attributable risks in cutaneous melanoma: Effects of proband and age. *J. Investig. Dermatol.* **2003**, *120*, 217–223. [CrossRef]
19. Wu, S.; Cho, E.; Li, W.-Q.; Qureshi, A.A. History of keratinocyte carcinoma and risk of melanoma: A prospective cohort study. *JNCI* **2017**, *109*, djw268. [CrossRef] [PubMed]
20. Wheless, L.; Black, J.; Alberg, A.J. Nonmelanoma skin cancer and the risk of second primary cancers: A systematic review. *Cancer Epidemiol. Prev. Biomark.* **2010**, *19*, 1686–1695. [CrossRef]
21. Fortes, C.; De Vries, E. Nonsolar occupational risk factors for cutaneous melanoma. *Int. J. Dermatol.* **2008**, *47*, 319–328. [CrossRef]
22. Wang, Y.; Cai, J.; Louie, D.C.; Wang, Z.J.; Lee, T.K. Incorporating clinical knowledge with constrained classifier chain into a multimodal deep network for melanoma detection. *Comput. Biol. Med.* **2021**, *137*, 104812. [CrossRef]
23. Bashkatov, A.N.; Zakharov, V.P.; Bucharskaya, A.B.; Borisova, E.G.; Khristoforova, Y.A.; Genina, E.A.; Tuchin, V.V. Malignant Tissue Optical Properties. In *Multimodal Optical Diagnostics of Cancer*; Tuchin, V.V., Popp, J., Zakharov, V., Eds.; Springer: Cham, Switzerland, 2020; pp. 3–106. [CrossRef]
24. Cordero, E.; Latka, I.; Matthäus, C.; Schie, I.W.; Popp, J. In-vivo Raman spectroscopy: From basics to applications. *J. Biomed. Opt.* **2018**, *23*, 071210. [CrossRef]
25. Jermyn, M.; Mercier, J.; Aubertin, K.; Desroches, J.; Urmey, K.; Karamchandiani, J.; Marple, E.; Guiot, M.-C.; Leblond, F.; Petrecca, K. Highly accurate detection of cancer in situ with intraoperative, label-free, multimodal optical spectroscopy. *Cancer Res.* **2017**, *77*, 3942–3950. [CrossRef]
26. Pacheco, A.G.; Krohling, R.A. The impact of patient clinical information on automated skin cancer detection. *Comput. Biol. Med.* **2020**, *116*, 103545. [CrossRef]
27. Zhao, J.; Zeng, H.; Kalia, S.; Lui, H. Incorporating patient demographics into Raman spectroscopy algorithm improves in vivo skin cancer diagnostic specificity. *Transl. Biophotonics* **2019**, *1*, 201900016. [CrossRef]
28. Kharazmi, P.; Kalia, S.; Lui, H.; Wang, Z.J.; Lee, T.K. A feature fusion system for basal cell carcinoma detection through data-driven feature learning and patient profile. *Ski. Res. Technol.* **2018**, *24*, 256–264. [CrossRef] [PubMed]
29. Bratchenko, I.A.; Bratchenko, L.A.; Moryatov, A.A.; Khristoforova, Y.A.; Artemyev, D.N.; Myakinin, O.O.; Kozlov, S.V.; Zakharov, V.P. In vivo diagnosis of skin cancer with a portable Raman spectroscopic device. *Exp. Dermatol.* **2021**, *30*, 652–663. [CrossRef]
30. Bratchenko, I.A.; Bratchenko, L.A.; Khristoforova, Y.A.; Moryatov, A.A.; Kozlov, S.V.; Zakharov, V.P. Classification of skin cancer using convolutional neural networks analysis of Raman spectra. *Comput. Methods Progr. Biomed.* **2022**, *219*, 106755. [CrossRef] [PubMed]
31. Pence, I.; Mahadevan-Jansen, A. Clinical instrumentation and applications of Raman spectroscopy. *Chem. Soc. Rev.* **2016**, *45*, 1958–1979. [CrossRef] [PubMed]
32. Bratchenko, I.A.; Khristoforova, Y.A.; Bratchenko, L.A.; Moryatov, A.A.; Kozlov, S.V.; Borisova, E.G.; Zakharov, V.P. Optical Biopsy of Amelanotic Melanoma with Raman and Autofluorescence Spectra Stimulated by 785 nm Laser Excitation. *J. Biomed. Photonics Eng.* **2021**, *7*, 020308. [CrossRef]
33. Kucheryavskiy, S. "Mdatools": Multivariate Data Analysis for Chemometrics. R Package Version 0.9.4. 2019. Available online: www.mdatools.com (accessed on 31 January 2022).
34. Chong, I.G.; Jun, C.H. Performance of some variable selection methods when multicollinearity is present. *Chemom. Intell. Lab Syst.* **2005**, *78*, 103–112. [CrossRef]
35. Robin, X.; Turck, N.; Hainard, A.; Tiberti, N.; Lisacek, F.; Sanche, J.-C.; Müller, M. pROC: An open-source package for R and S+ to analyze and compare ROC curves. *BMC Bioinform.* **2011**, *12*, 77. [CrossRef]
36. Hanley, J.A.; McNeil, B.J. A method of comparing the areas under receiver operating characteristic curves derived from the same cases. *Radiology* **1983**, *148*, 839–843. [CrossRef]
37. Kaprin, A.D.; Sratinskiy, V.V.; Shakhzadova, A.O. (Eds.) *Malignant Tumors in Russia in 2020 (Morbidity and Mortality)*; National Medical Research Radiological Centre of the Ministry of Health of the Russian Federation: Moscow, Russian Federation, 2021;

p. 252. ISBN 978-5-85502-268-1. Available online: https://oncology-association.ru/wp-content/uploads/2021/11/zis-2020-elektronnaya-versiya.pdf. (In Russian)

38. Kvaskoff, M.; Pandeya, N.; Green, A.C.; Perry, S.; Baxter, C.; Davis, M.B.; Mortimore, R.; Westacott, L.; Wood, D.; Triscott, J.; et al. Site-specific determinants of cutaneous melanoma: A case–case comparison of patients with tumors arising on the head or trunk. *Cancer Epidemiol. Prev. Biomark.* **2013**, *22*, 2222–2231. [CrossRef]
39. Khristoforova, Y.A.; Bratchenko, I.A.; Myakinin, O.O.; Artemyev, D.N.; Moryatov, A.A.; Orlov, A.E.; Kozlov, S.V.; Zakharov, V.P. Portable spectroscopic system for in vivo skin neoplasms diagnostics by Raman and autofluorescence analysis. *J. Biophotonics* **2019**, *12*, e201800400. [CrossRef] [PubMed]
40. Haenssle, H.; Fink, C.; Toberer, F.; Winkler, J.; Stolz, W.; Deinlein, T.; Hofmann-Wellenhof, R.; Lallas, A.; Emmert, S.; Buhl, T.; et al. Man against machine reloaded: Performance of a market-approved convolutional neural network in classifying a broad spectrum of skin lesions in comparison with 96 dermatologists working under less artificial conditions. *Ann. Oncol.* **2020**, *31*, 137–143. [CrossRef] [PubMed]
41. Estava, A.; Kuprel, B.; Novoa, R.; Ko, J.; Swetter, S.M.; Blau, H.M.; Thrun, S. Dermatologist level classification of skin cancer with deep neural networks. *Nature* **2017**, *542*, 115–118. [CrossRef] [PubMed]
42. Alam, T.M.; Shaukat, K.; Khan, W.A.; Hameed, I.A.; Almuqren, L.A.; Raza, M.A.; Luo, S. An Efficient Deep Learning-Based Skin Cancer Classifier for an Imbalanced Dataset. *Diagnostics* **2022**, *12*, 2115. [CrossRef]

*diagnostics*

MDPI

*Article*

# Study of the Natural Crystalline Lens Characteristics Using Dual-Energy Computed Tomography

**Jeffrey R. Sachs [1], Javier A. Nahmias [2], Kevin D. Hiatt [1], James G. Bomar [1], Thomas G. West [1], Paul M. Bunch [1], Marc D. Benayoun [1], Chris Lack [1] and Atalie C. Thompson [2,*]**

1 Department of Radiology, Atrium Health Wake Forest Baptist Medical Center, 1 Medical Center Boulevard, Winston-Salem, NC 27157, USA

2 Department of Surgical Ophthalmology, Atrium Health Wake Forest Baptist Medical Center, 1 Medical Center Boulevard, Winston-Salem, NC 27157, USA

* Correspondence: atathomp@wakehealth.edu; Tel.: +1-(650)-868-8050; Fax: +1-(336)-716-9334

**Abstract:** There is a paucity of radiologic literature regarding age-related cataract, and little is known about any differences in the imaging appearance of the natural crystalline lens on computed tomography (CT) exams among different demographic groups. In this retrospective review of 198 eyes in 103 adults who underwent dual-energy computed tomography (DECT) exams of the head, regions of interest spanning 3–5 mm were placed over the center of the lens, and the x-ray attenuation of each lens was recorded in Hounsfield Units (HU) at 3 energy levels: 40 keV, 70 keV, and 190 keV. Generalized estimating equations (GEEs) were used to assess the association of clinical or demographic data with lens attenuation. The mean HU values were significantly lower for the older vs. younger group at 40 keV (GEE $p$-value = 0.022), but there was no significant difference at higher energy levels ($p > 0.05$). Mean HU values were significantly higher for females vs. males and non-whites vs. non-Hispanic whites at all 3 energy levels in bivariate and multivariable analyses (all $p$-value < 0.05). There was no significant association between lens attenuation and either diabetes or smoking status. The crystalline lens of females and non-whites had higher attenuation on DECT which may suggest higher density or increased concentration of materials like calcium and increased potential for cataract formation. Given the large scope of cataracts as a cause of visual impairment and the racial disparities that exist in its detection and treatment, further investigation into the role of opportunistic imaging to detect cataract formation is warranted.

**Keywords:** dual-energy computed tomography; natural crystalline lens; sex; race

**Citation:** Sachs, J.R.; Nahmias, J.A.; Hiatt, K.D.; Bomar, J.G.; West, T.G.; Bunch, P.M.; Benayoun, M.D.; Lack, C.; Thompson, A.C. Study of the Natural Crystalline Lens Characteristics Using Dual-Energy Computed Tomography. *Diagnostics* **2022**, *12*, 2857. https://doi.org/10.3390/diagnostics12112857

Academic Editor: Viktor Dremin

Received: 22 September 2022
Accepted: 15 November 2022
Published: 18 November 2022

**Publisher's Note:** MDPI stays neutral with regard to jurisdictional claims in published maps and institutional affiliations.

**Copyright:** © 2022 by the authors. Licensee MDPI, Basel, Switzerland. This article is an open access article distributed under the terms and conditions of the Creative Commons Attribution (CC BY) license (https://creativecommons.org/licenses/by/4.0/).

## 1. Introduction

Cataracts are the leading cause of blindness worldwide and affect over 24 million Americans [1]. Of the subtypes of age-related cataracts, nuclear sclerotic cataracts are the most common [2], and are thought to result from cumulative lifetime exposure to a range of insults to the ocular lens including ultraviolet light, ocular trauma, ocular surgery, corticosteroid use, radiation exposure, smoking, and diabetes mellitus.

Currently, cataracts are almost universally a clinical diagnosis, made by confirming opacification or discoloration of the lens during slit lamp examination by a trained specialist in optometry or ophthalmology. However, barriers to access of basic healthcare services have caused many patients to seek acute medical care through the emergency room, which provides an opportunity for incidental discovery of comorbid chronic conditions, such as cataracts [3].

Computed tomography (CT) scans are one of the most common medical tests performed in the emergency room [4]. CT has proven useful in diagnosing traumatic cataracts [5–8], wherein the lens becomes hypoattenuating due to increased fluid content. However, there is a striking paucity of radiologic literature on the imaging characteristics of sage-related cataracts, which are far more common than traumatic cataracts [2]. Moreover, little is known about differences in the characteristics of the natural lens on CT imaging among different demographic groups.

One reason for this lack of knowledge is that when using conventional single-energy CT techniques, the lenses are bland, relatively homogeneous structures that garner little attention unless they are displaced, dysmorphic, or absent. Moreover, since the natural lens is a radiosensitive organ at particular risk for radiation effects due to its superficial location in the human body, intentional imaging of the lens during radiologic image acquisition is not generally recommended [9]. In clinical practice, however, the lens is often incidentally included in the CT scan field of view. This circumstance presents the radiologist with an opportunity to assess the lens for pathology such as cataracts.

In recent years, dual-energy CT (DECT) has risen to the forefront of CT imaging acquisition due to its ability to capitalize on the differences in energy-dependent x-ray absorption of different materials within the patient by using low- and high-energy x-ray spectra. DECT techniques have numerous recently described applications in the practice of neuroradiology [10]. Low kilo–electron volt (keV) (e.g., 40 keV) virtual monoenergetic imaging (VMI) techniques allow for improved contrast-to-noise ratio among soft tissues of similar attenuation, even in the absence of iodinated contrast media [11]. High keV (e.g., 190 keV) VMI have been used to describe unique attenuation patterns of silicone oil which can be found in the eye after certain retinal surgeries [12,13].

Given the ability of DECT to improve discrimination between a material of high atomic number (such as calcium or iodine) and a material of low atomic number (such as hemorrhage) [14], we hypothesized that it would be useful in detecting differences in lens composition that could be related to early cataract formation. The lens is made of proteins which are low in atomic number. If the cataractic lens were to calcify, then one would expect that the attenuation of the lens would increase fairly dramatically over time, especially at 40 keV, due to the combined impact of low energy incident photons and high atomic number on the photoelectric effect, and subsequently on x-ray beam attenuation. Although total lens calcification is rare [15], lens calcium content has been shown to correlate with the degree of opacification of cataractous lenses [16]. Impaired intracellular Ca+2 signaling in lens epithelial cells is known to play an intrinsic role in both cortical and nuclear cataractogenesis [17]. Moreover, we might expect nuclear cataracts to have an increased density due to the pathologic aggregation and compaction of the proteins of the nuclear fibers which has been observed on histology [2]. However, to our knowledge, no study has investigated the characteristics of the natural crystalline lens on DECT or whether there are differences in attenuation of the lens in different demographic groups or at different levels of energy.

In this retrospective study, we examined whether there is a relationship between possible demographic and clinical risk factors for cataract formation, such as age, race, sex, diabetes and smoking status, with the X-ray attenuation of the natural crystalline lens on DECT of the head at 40, 70, and 190 keV VMI. Such imaging analyses of the crystalline lens may provide critical foundational understanding of the changes in the lens that one might observe on DECT as well as help identify groups at potential risk for cataract formation.

## 2. Materials and Methods

### 2.1. Subjects

For this retrospective, Health Insurance Portability and Accountability Act-compliant, institutional review board-approved study, a local institutional radiology database (syngo.via, Siemens Healthineers) was queried with the goal of identifying eligible lenses for study inclusion. Screening eligibility criteria were (1) age ≥18 years and (2) received non-contrast DECT of the head at our institution between July and December 2020. A younger adult (age ≥18 and <30 years) and older adult cohort (age >70 years) were collected to facilitate assessment of any age-related differences in the lens characteristics. Patients were excluded if there was evidence of acute orbital trauma (including periorbital hematoma, orbital hemorrhage, and orbital fractures). Pseudophakic eyes were not included in the analysis. Using these criteria, a total of 198 eyes in 103 adult subjects were identified. Demographic or clinical risk factors for cataract such as age, sex, race/ethnicity, smoking status, and history of diabetes were collected from the electronic medical record. Documentation of known cataract status was also collected when available.

### 2.2. Image Acquisition

All non-contrast DECT examinations of the head were performed utilizing a dual-source system (SOMATOM Drive or Flash; Siemens Healthineers, Erlangen, Germany). The DECT acquisition parameters were as follows: 80 kVp/Sn140 kVp acquisitions, Quality Reference mAs of 400/200, pitch of 0.70, rotation time of 0.5 s, and with automated tube current modulation, CareDose 4D.

### 2.3. Image Analysis

All dual-energy post-processing and region of interest (ROI) analysis was performed in syngo.via (Siemens Healthineers) using the Monoenergetic+ application. Measurements were obtained across all subjects using an image slice thickness of 1 mm. Two-dimensional circular ROIs were drawn by a trained radiology resident or neuroradiology fellow in the center of the native lenses. Care was taken to identify the center of the lens and avoid any streak or beam hardening artifacts. An attending neuroradiologist with 5 years subspecialty experience reviewed and optimized the ROI position as needed. All ROIs were sized to be between 3–5 mm$^2$. A representative example of how ROI measurements were obtained is given in Figure 1. After the initial ROI was placed, the ROI location did not change while HU attenuation measurements were recorded at each of the 40 keV, 70 keV, and 190 keV energy levels. The mean attenuation and standard deviation (SD) were recorded in Hounsfield Units (HU) for each ROI at each energy level.

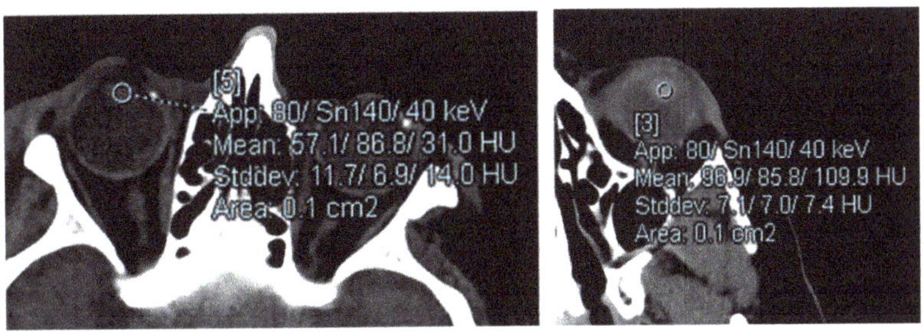


**Figure 1.** A representative example of how ROI measurements were obtained for the study. All measurements were obtained in the axial plane using 1 mm thick slices. ROI data was obtained using the default syngo.via Monoenergetic+ application. All scans were non-contrast dual-source, dual-energy CT exams. See the text Section 2.2. for the DECT acquisition parameters.

### 2.4. Statistical Analyses

Descriptive statistics were used to describe the clinical and demographic characteristics of the population. Pearson's correlation was used to compare the attenuation in Hounsfield units between right and left eyes among subjects with both natural lenses present on DECT. In order to account for the correlation of both eyes within a given subject, separate generalized estimating equations (GEEs) were constructed to assess the bivariate relationship between each of the demographic or clinical variables and the relative attenuation of the lens at each energy level (40, 70, and 190 keV). Factors that were significant in bivariate analyses ($p$-value < 0.05) were entered into multivariable GEE models of the HU for each energy level. All analyses were performed with Stata (version 17.0, StataCorp, College Station, TX, USA). A $p$-value < 0.05 was considered statistically significant.

## 3. Results

### 3.1. Subjects

Included subjects were sub-categorized into an older adult (N = 53; mean age 81.4 ± 5.7 years; range 73–101) and younger adult cohort (N = 50; mean age 22.66 ± 2.93 years; range 18–27). Approximately 46.6% of subjects were female, 38.8% self-identified as non-Hispanic white, 22.3% had diabetes, and 37.9% were never smokers (Table 1). Among the older adults, only 19 subjects (29 eyes) had ophthalmic examination data of the natural lens in their electronic healthcare record so that cataract status could be determined.

**Table 1.** Clinical and demographic characteristics of 103 adult subjects who underwent Dual Energy Computed Tomography (DE-CT).

| Characteristics | N (%) |
|---|---|
| **Age** | |
| Older Adults (73–101 years) | 53 (51.5%) |
| Young Adults (18–27 years) | 50 (48.5%) |
| **Sex** | |
| Female | 48 (46.6%) |
| Male | 55 (53.4%) |
| **Race/Ethnicity** | |
| Non-White | 40 (38.8%) |
| Non-Hispanic White | 63 (61.2%) |
| **Smoking Status** | |
| Ever Smoker | 39 (37.9%) |
| Never-smoker | 64 (62.1%) |
| **Diabetes** | |
| Yes | 23 (22.3%) |
| No | 80 (77.7%) |

### 3.2. Image Analysis

Ninety-five subjects had a natural crystalline lens present in both eyes. There was a strong and significant positive correlation between the HU attenuation values in the right and left eye at each of the three energy levels (all $p < 0.0001$) (Table 2).

**Table 2.** Correlation of right and left eye attenuation measurements on Dual-Energy Computed Tomography in 95 pairs of eyes in 95 subjects.

| | **Mean ± Standard Deviation, Hounsfeld Units in keV** | **Mean ± Standard Deviation, Hounsfeld Units in keV** | | |
|---|---|---|---|---|
| **DECT Energy Level** | Right Eye | LEFT EYE | Pearson's Correlation | *p*-Value |
| 40 keV | 73.5 ± 12.6 | 74.0 ± 14.1 | 0.4061 | **<0.0001** |
| 70 keV | 76.3 ± 8.0 | 76.6 ± 7.9 | 0.6706 | **<0.0001** |
| 190 keV | 77.5 ± 8.7 | 77.7 ± 8.8 | 0.6926 | **<0.0001** |

DECT = Dual-Energy Computed Tomography. Bolded *p*-values are statistically significant ($p < 0.05$).

Table 3 displays the bivariate association between clinical and demographic characteristics and attenuation of the natural crystalline lens in the full cohort (198 eyes in 103 subjects). At 40 keV, the older adults had a significantly lower average HU than younger adults (70.88 ± 13.63 vs. 75.70 ± 13.21; $p = 0.020$), but there was no significant difference in the measurements for older vs. younger adults at the 70 keV or 190 keV energy levels (both $p > 0.05$). However, in the subgroup of older adults with clinical documentation of a cataract (N = 19 subjects with 29 eyes), the Hounsfield unit values were higher, measuring 74.65 ± 14.57) at 40 keV, 79.05 ± 7.58 at 70 keV, and 80.85 ± 9.29 at 190 keV.

**Table 3.** Bivariate analysis of clinical and demographic characteristics associated with natural crystalline lens attenuation measured at three energy levels on Dual Energy Computed Tomography (DE-CT) (N = 198 eyes in 103 subjects).

| Characteristics | Hounsfield Units (HU) in keV | | |
|---|---|---|---|
| | 40 keV | 70 keV | 190 keV |
| **Age** | | | |
| Older Adults | 70.88 ± 13.63 | 75.18 ± 7.77 | 76.91 ± 9.26 |
| Young Adults | 75.70 ± 13.21 | 77.64 ± 7.78 | 78.48 ± 8.17 |
| GEE *p*-value | **0.022** | 0.080 | 0.381 |
| **Sex** | | | |
| Female | 76.41 ± 13.28 | 79.0 ± 7.37 | 79.99 ± 7.91 |
| Male | 70.68 ± 13.37 | 74.24 ± 7.61 | 75.76 ± 8.97 |
| GEE *p*-value | **0.010** | **<0.0001** | **0.005** |
| **Race/Ethnicity** | | | |
| Non-White | 77.63 ± 13.12 | 79.98 ± 7.86 | 81.03 ± 8.12 |
| Non-Hispanic White | 70.51 ± 13.22 | 74.11 ± 6.96 | 75.54 ± 8.48 |
| GEE *p*-value | **0.001** | **<0.0001** | **0.001** |
| **Smoking Status** | | | |
| Ever Smoker | 71.81 ± 14.92 | 75.30 ± 8.04 | 76.69 ± 9.71 |
| Non-smoker | 74.23 ± 12.71 | 77.12 ± 7.68 | 78.32 ± 8.07 |
| GEE *p*-value | 0.275 | 0.224 | 0.361 |
| **Diabetes** | | | |
| Yes | 70.83 ± 13.01 | 75.96 ± 6.88 | 77.98 ± 9.59 |
| No | 73.98 ± 13.72 | 76.55 ± 8.11 | 77.63 ± 8.53 |
| GEE *p*-value | 0.168 | 0.810 | 0.588 |

Generalized estimating equations = GEE. Bolded *p*-values are statistically significant ($p < 0.05$).

At all three energy levels, the mean HU attenuation values were significantly higher for females vs. males (40 keV: 76.4 ± 13.3 vs. 70.7 ± 13.4; 70 keV: 79.0 ± 7.4 vs. 74.2 ± 7.6; 190 keV: 80 ± 7.9 vs. 75.8 ± 9; all $p <= 0.01$) and for non-whites vs. non-Hispanic whites (40 keV: 77.6 ± 13.1 vs. 70.5 ± 13.2; 70 keV: 80 ± 7.9 vs. 74.1 ± 7.0; 190 keV: 81.0 ± 8.1 vs. 75.5 ± 8.5; all $p <= 0.001$). No significant association was detected between lens attenuation and either diabetes or smoking status (all $p > 0.05$).

In multivariable analyses, both female sex (all $p < 0.01$) and nonwhite race/ethnicity (all $p < 0.01$) but not age-group ($p > 0.05$) remained significant independent predictors of lens attenuation at all 3 energy levels (Table 4).

**Table 4.** Multivariable generalized estimating equations of characteristics associated with natural crystalline lens attenuation measured at three energy levels on Dual Energy Computed Tomography (DE-CT) (N = 198 eyes in 103 subjects).

| Characteristics | Hounsfield Units (HU) in keV | | |
|---|---|---|---|
| | 40 keV<br>Beta (95% CI) | 70 keV<br>Beta (95% CI) | 190 keV<br>Beta (95% CI) |
| **Age** | | | |
| Older Adults | 0.024<br>(0.0003, 1.63) | 0.272<br>(0.022, 3.34) | 0.789<br>(0.406, 15.33) |
| Young Adults | Reference | Reference | Reference |
| GEE *p*-value | 0.083 | 0.309 | 0.876 |
| **Sex** | | | |
| Female | 374.5<br>(6.54, 21434.22) | 124.42<br>(11.3, 1369.9) | 71.06<br>(4.17, 1210.02) |
| Male | Reference | Reference | Reference |
| GEE *p*-value | **0.004** | **<0.0001** | **0.003** |
| **Race/Ethnicity** | | | |
| Non-White | 403.47<br>(5.367, 30333.49) | 208.07<br>(16.04, 2698.38) | 179.92<br>(8.70, 3718.84) |
| Non-Hispanic White | Reference | Reference | Reference |
| GEE *p*-value | **0.006** | **<0.0001** | **0.001** |

GEE = Generalized estimating equations; CI = confidence interval. Bolded *p*-values are statistically significant ($p < 0.05$).

## 4. Discussion

To our knowledge, this study is the first to demonstrate the ability of DECT to detect differences in the attenuation of the crystalline lens on DECT among demographic and clinical groups that may be at increased risk for cataract formation.

We found that the crystalline lens of females and non-white subjects had significantly higher attenuation on DECT at all energy levels. This finding may suggest a higher density lens or increased concentration of materials like calcium which could be related to lens opacification from early cataract formation [16]. The Salisbury Eye Evaluation Study found that African Americans had higher rates of cortical opacity and progression of cortical cataracts [18]. The Age-Related Eye Disease study similarly found that both females and non-whites were at greater risk of cortical cataract formation [19]. One hypothesis in females is that age-related withdrawal of estrogen may play a role in the progression cataracts through the loss of its anti-oxidative effects [20]. Women also have a higher prevalence of osteoporosis which has been associated with increased risk for cataract, presumably through common pathways of impaired calcium homeostasis [21]. Such impaired calcium signaling in lens epithelial cells can result in increased cytosolic calcium concentration [22] which predisposes to cortical cataracts in particular. For example, one study demonstrated that the total calcium in lenses with cortical cataracts measured four times higher than in clear lenses [23]. Cortical cataracts can also have discrete calcium deposits which would have substantially higher attenuation on DECT. Moreover, impaired calcium signaling can also lead to opacification of the nuclear portion of the lens resulting in a mixed cataract type [17]. Interestingly, the subset of older adults with a documented cataract also had higher HU values. However, the retrospective design of the study limited more direct assessment of the relationship between these characteristics and cataract formation since ophthalmic examination data was not available in a majority of patients. Nevertheless, the study demonstrated that DECT has the potential to identify differences in lens attenuation among groups at risk for cataract. Such findings could form the rationale for a future prospective study in which patients undergoing DECT of the head are recruited for ophthalmology examination to determine whether a clinically and visually significant opacification of the lens is present.

To our surprise, no significant association between lens attenuation and either diabetes or smoking status was detected (all $p > 0.05$). High blood glucose in the setting of poorly controlled diabetes can lead to generation of polyols that result in increased osmotic stress in the lens fibers causing them to swell and rupture [24]. Given previous studies on traumatic cataracts and the low HU attenuation from increased water concentration, one might have expected diabetic patients to have a lower HU attenuation. The lack of an association here could be related to the fact that severity of diabetes was not able to be determined since HbA1c was not consistently available. Future studies could collect additional lab criteria to stratify by diabetic control. Similarly, prior studies have associated smoking with an increased risk of nuclear sclerotic type cataracts due to increased oxidative stress [25,26]. It is possible that smoking data from the medical record is less accurate than questionnaire data which could be collected in a future prospective study.

The older adult cohort in general trended toward a lower attenuation of the lens at 40 keV, but this relationship was not significant in multivariable analyses or at higher energy levels. It is possible that DECT did not identify a Hounsfield-unit based threshold to distinguish lens age because different subtypes of cataract may result in different alterations in lens attenuation. For example, some cataracts might decrease lens attenuation if there is an increase in fluid whereas others may increase lens attenuation if there is an increase in deposition of calcium or other high-density materials. Future studies could consider whether specific subtypes of cataract have different degrees of lens attenuation on DECT.

Although an ROI based methodology to confirm cataract by DECT is not yet possible, the lens remains a potentially attractive target for artificial intelligence-based segmentation given its simple shape, near-uniform size, and the high level of contrast between the lens itself and the surrounding fluid of the aqueous and vitreous humor [27]. Larger DECT imaging datasets pooled across multiple institutions may make possible the training and development of deep learning algorithms for predicting cataract risk. Additionally, while obtaining the study data we noticed occasional patients that demonstrated highly variable attenuation values in the lenses that produced a "speckled" pattern of attenuation (Figure 2). Such a pattern suggests the usefulness of texture analysis for identification of cataracts, where differences in spatial heterogeneity within the ROI of a cataract lens can be distinguished from a non-cataract lens, despite having, for example, similar overall mean HU attenuation values. In fact, prior studies have reported greater than 90% accuracy in identifying cataracts from eye photographs when using lens ROI uniformity (a texture analysis metric) as a training feature for a nearest-neighbor classifier [28], though to the best of our knowledge this has not yet been reproduced with DECT images.

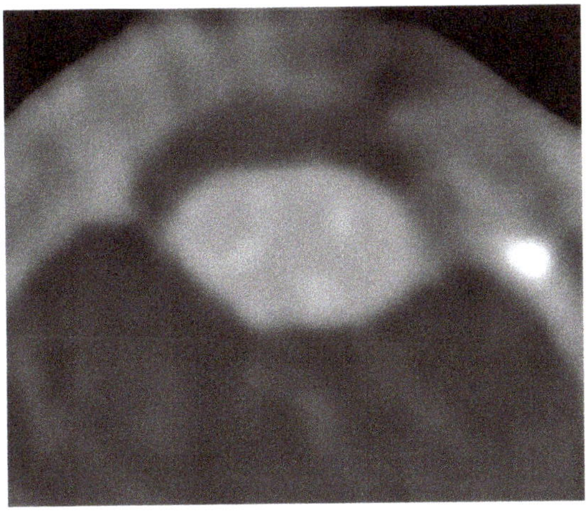

**Figure 2.** Axial non-contrast DECT image zoomed for lens detail. An 84-year-old lens demonstrates a "speckled" appearance—with scattered punctate foci of increased attenuation. This pattern of spatial heterogeneity raises the question of whether texture analysis may be a more useful method of assessing for cataract at CT.

A majority of patients in our study did not have historic ophthalmic data, suggesting there could be gaps in access to ophthalmic care that could be potentially addressed during an emergency department visit. Racial disparities in access to cataract surgery are well-documented, especially among black patients [29]. Inequities in the proportion of patients that carry health insurance affects access to routine health care and utilization [30,31]. Racial minorities are also known to be less likely than white patients to have a primary care provider and are more likely to rely on the Emergency Department for routine care needs [32–34]. As such, radiologists could have an opportunity to help improve healthcare disparities through invention of novel opportunistic imaging screening techniques (i.e., screening an organ for pathology when that organ is incidentally imaged as part of a study obtained for a separate indication). While traditional cataract evaluation will remain the gold-standard for cataract diagnosis, future studies should investigate whether patients presenting to the ED who undergo DE-CT could benefit from opportunistic detection of cataract and development of a care pathway that refers such patients to ophthalmology for confirmation and treatment.

Our study has several limitations. In order to allow for examination of possible age-related differences in the lens, two cohorts of older and younger adults were collected. Thus, the study findings are not generalizable to middle-aged patients. Achieving accurate lens cortex attenuation values is problematic due to the concern for volume averaging with adjacent fluid filled structures. However, if eyes had cortical changes overlying the nucleus these would have been detected within the ROI. Additional ophthalmic clinical data as to the presence or grade of cataract were only available in a subset of older adults. Based on the findings in this study, a future prospective study could be designed in which patients are referred to an ophthalmologist for clinical exam of the lens following incidental imaging of the crystalline lens on DECT during an emergency room encounter.

Another potential limitation of the study is the presence of beam hardening artifacts at 40 keV VMI. In our study, the authors took great care to avoid beam hardening artifacts emanating from the bony orbit; if such artifacts had been included in the ROI analysis, they would have led to spurious increases or decreases in HU values. Photon counting CT holds promise to both improve spectral resolution and reduce beam hardening artifacts in the future [35].

Our study was performed on one of two distinct types of DECT scanners, both dual-source systems manufactured by Siemens, and with nearly identical protocols. These results may not necessarily be generalizable to all commercially available forms of spectral CT imaging, though the underlying physics should remain the same. Moreover, several factors can affect the accuracy of the HU values, including the aforementioned beam hardening artifacts, spectral energy, convolution kernel, and patient positioning [36]. CT scanners are calibrated such that the HU value for pure water does not deviate more than 2 HU from the reference value of 0 [37], but the possibility exists that small differences in scanner calibration in conjunction with a combination of the above factors could impact the measurements taken.

## 5. Conclusions

Our study found that non-white and female patients had significantly higher lens attenuation values on DECT, which may suggest increased lens density, the presence of high-density materials like calcium, or a predisposition for cataracts. Future collection of ophthalmic examination data in patients with DECT findings will provide useful diagnostic confirmation of the clinical relevance of these findings. Given the large scope of cataracts as a cause of visual impairment and the racial disparities that exist in its detection and treatment, further investigation into the role of opportunistic DECT imaging to detect cataract formation is warranted. As these technologies continue to advance, incorporation of photon counting CT and artificial intelligence-based textural analysis may improve the utility of such radiologic studies in the clinical care of patients with incidental cataracts.

**Author Contributions:** Conceptualization, J.R.S., M.D.B., T.G.W., P.M.B., C.L., A.C.T.; Methodology, J.R.S., A.C.T., P.M.B., C.L., M.D.B., K.D.H., J.G.B., T.G.W.; Software, J.R.S., A.C.T.; Validation, J.R.S., A.C.T.; Formal analysis, J.R.S., A.C.T.; Investigation, J.R.S., K.D.H., J.G.B., A.C.T., J.A.N.; Resources, J.R.S., A.C.T.; Data curation, J.R.S., A.C.T., J.A.N.; Writing—original draft preparation, J.R.S., A.C.T., J.A.N.; Writing—review and editing, J.R.S., A.C.T., J.A.N., K.D.H., T.G.W., P.M.B., M.D.B., C.L., J.G.B.; Visualization, J.R.S.; Supervision, J.R.S., A.C.T.; Project administration, not applicable; Funding acquisition, not applicable. All authors have read and agreed to the published version of the manuscript.

**Funding:** This research received no external funding. Dr. Thompson receives support from the NIH/NEI (K23-EY030897) and the Wake Forest Older Americans Independence Center (P30-AG021332). Dr. Bunch received support as an AUR GERRAF fellow.

**Institutional Review Board Statement:** The study was conducted in accordance with the Declaration of Helsinki, and approved by the Institutional Review Board Wake Forest Baptist Medical Center (protocol code IRB00068983 and approved 10/5/2020).

**Informed Consent Statement:** A waiver of informed consent was granted by the Wake Forest Baptist Health Medical Center Institutional Review Board due to the retrospective nature of this study. All collected data were de-identified and study results were published in aggregate.

**Data Availability Statement:** Restrictions apply to the availability of these data. Data were obtained from the electronic healthcare record at Atrium-Wake Forest Baptist Health and a de-identified dataset may be available with the permission of Atrium-Wake Forest Baptist Health's legal team.

**Conflicts of Interest:** The authors declare no conflict of interest.

## References


1. Congdon, N.; Vingerling, J.R.; Klein, B.E.K.; West, S.; Friedman, D.; Kempen, J.; O'Colmain, B.; Wu, S.-Y.; Taylor, H.R. Eye Diseases Prevalence Research Group Prevalence of Cataract and Pseudophakia/Aphakia Among Adults in theUnited States. *Arch. Ophthalmol.* **2004**, *122*, 487–494. [CrossRef] [PubMed]
2. Aliancy, J.F.; Mamalis, N. Crystalline Lens and Cataract. 2017 Aug 15. In *Webvision: The Organization of the Retina and Visual System;* Kolb, H., Fernandez, E., Nelson, R., Eds.; University of Utah Health Sciences Center: Salt Lake City, UT, USA, 1995. Available online: https://www.ncbi.nlm.nih.gov/books/NBK476171/ (accessed on 15 August 2022).
3. Rust, G.; Ye, J.; Baltrus, P.; Daniels, E.; Adesunloye, B.; Fryer, G.E. Practical Barriers to Timely Primary Care Access. *Arch. Intern. Med.* **2008**, *168*, 1705–1710. [CrossRef] [PubMed]
4. Kocher, K.E.; Meurer, W.; Fazel, R.; Scott, P.A.; Krumholz, H.M.; Nallamothu, B.K. National Trends in Use of Computed Tomography in the Emergency Department. *Ann. Emerg. Med.* **2011**, *58*, 452–462. [CrossRef]
5. Almog, Y.; Reider-Groswasser, I.; Goldstein, M.; Lazar, M.; Segev, Y.; Geyer, O. The Disappearing Lens: Failure of CT to Image the Lens in Traumatic Intumescent Cataract. *J. Comput. Assist. Tomogr.* **1999**, *23*, 354–356. [CrossRef]
6. Segev, Y.; Goldstein, M.; Lazar, M.; Reider-Groswasser, I. CT appearance of a traumatic cataract. *AJNR Am. J. Neuroradiol.* **1995**, *16*, 1174–1175.
7. Taslakian, B.; Hourani, R. Don't forget to report "simple" finding on CT: The hypodense eye lens. *Eur. J. Pediatr.* **2013**, *172*, 131–132. [CrossRef] [PubMed]
8. Boorstein, J.M.; Titelbaum, D.S.; Patel, Y.; Wong, K.T.; Grossman, R. CT diagnosis of unsuspected traumatic cataracts in patients with complicated eye injuries: Significance of attenuation value of the lens. *Am. J. Roentgenol.* **1995**, *164*, 181–184. [CrossRef] [PubMed]
9. Yuan, M.-K.; Tsai, D.-C.; Chang, S.-C.; Yuan, M.-C.; Chang, S.-J.; Chen, H.-W.; Leu, H.-B. The Risk of Cataract Associated With Repeated Head and Neck CT Studies: A Nationwide Population-Based Study. *AJR Am. J. Roentgenol.* **2013**, *201*, 626–630. [CrossRef]
10. Sachs, J.R.; West, T.G.; Lack, C.M.; Perry, B.; Zapadka, M.E. How to Incorporate Dual-Energy Computed Tomography Into Your Neuroradiology Practice: Questions and Answers. *J. Comput. Assist. Tomogr.* **2018**, *42*, 824–830. [CrossRef]
11. Pomerantz, S.R.; Kamalian, S.; Zhang, D.; Gupta, R.; Rapalino, O.; Sahani, D.V.; Lev, M.H. Virtual Monochromatic Reconstruction of Dual-Energy Unenhanced Head CT at 65–75 keV Maximizes Image Quality Compared with Conventional Polychromatic CT. *Radiology* **2013**, *266*, 318–325. [CrossRef]
12. Bunch, P.M.; Lack, C.M.; West, T.G.; Tan, J.; Sachs, J.R. Accuracy of 190-keV Virtual Monoenergetic Dual-Energy CT for Determining the Presence or Absence of Intravitreal Silicone Oil. *Am. J. Roentgenol.* **2021**, *216*, 1040–1045. [CrossRef] [PubMed]
13. Sachs, J.R.; Lack, C.M.; West, T.G.; Bunch, P.M. Unique Characteristics of Intravitreal Silicone Oil on Dual-Energy Computed To-mography. *J. Comput. Assist. Tomogr.* **2020**, *44*, 370–373. [CrossRef] [PubMed]
14. Gupta, R.; Phan, C.M.; Leidecker, C.; Brady, T.J.; Hirsch, J.A.; Nogueira, R.G.; Yoo, A.J. Evaluation of Dual-Energy CT for Differentiating Intracerebral Hemorrhage from Iodinated Contrast Material Staining. *Radiology* **2010**, *257*, 205–211. [CrossRef] [PubMed]
15. Albert, D.M.; Miller, J.W.; Azar, D.T.; Young, E.H. Cataractogenesis in the Adult. In *Albert & Jakobiec's Principles and Practice of Ophthal-Mology;* Saunders Elsevier Inc.: Philadelphia, PA, USA, 2008.
16. Hightower, K.R.; Hind, D. Cytotoxic effects of calcium on sodium-potassium transport in the mammalian lens. *Curr. Eye Res.* **1982**, *2*, 239–246. [CrossRef]
17. Gosak, M.; Markovič, R.; Fajmut, A.; Marhl, M.; Hawlina, M.; Andjelić, S. The Analysis of Intracellular and Intercellular Calcium Signaling in Human Anterior Lens Capsule Epithelial Cells with Regard to Different Types and Stages of the Cataract. *PLoS ONE* **2015**, *10*, e0143781. [CrossRef]
18. Storey, P.; Munoz, B.; Friedman, D.; West, S. Racial Differences in Lens Opacity Incidence and Progression: The Salisbury Eye Evaluation (SEE) Study. *Investig. Ophthalmol. Vis. Sci.* **2013**, *54*, 3010–3018. [CrossRef]
19. Chang, J.R.; Koo, E.; Agrón, E.; Hallak, J.; Clemons, T.; Azar, D.; Sperduto, R.D.; Ferris, F.L.; Chew, E.Y. Risk Factors Associated with Incident Cataracts and Cataract Surgery in the Age-Related Eye Disease Study (AREDS): AREDS Report Number 32. *Ophthalmology* **2011**, *118*, 2113–2119. [CrossRef]
20. Zetterberg, M.; Celojevic, D. Gender and Cataract—The Role of Estrogen. *Curr. Eye Res.* **2014**, *40*, 176–190. [CrossRef]
21. Nemet, A.Y.; Hanhart, J.; Kaiserman, I.; Vinker, S. Are cataracts associated with osteoporosis? *Clin. Ophthalmol.* **2013**, *7*, 2079–2084. [CrossRef]
22. Mukai, K.; Matsushima, H.; Ishii, Y.; Obara, Y. Effects of calcium on lens epithelial cells in rabbits. *Nippon Ganka Gakkai Zasshi* **2006**, *110*, 361–369.
23. Tang, D.; Borchman, D.; Yappert, M.C.; Vrensen, G.F.J.M.; Rasi, V. Influence of Age, Diabetes, and Cataract on Calcium, Lipid-Calcium, and Protein-Calcium Relationships in Human Lenses. *Investig. Opthalmol. Vis. Sci.* **2003**, *44*, 2059–2066. [CrossRef] [PubMed]
24. Pollreisz, A.; Schmidt-Erfurth, U. Diabetic Cataract—Pathogenesis, Epidemiology and Treatment. *J. Ophthalmol.* **2010**, *2010*, 608751. [CrossRef] [PubMed]
25. Ye, J.; He, J.; Wang, C.; Wu, H.; Shi, X.; Zhang, H.; Xie, J.; Lee, S.Y. Smoking and Risk of Age-Related Cataract: A Meta-Analysis. *Investig. Ophthalmol. Vis. Sci.* **2012**, *53*, 3885–3895. [CrossRef] [PubMed]

26. Cheng, A.C.; Pang, C.P.; Leung, A.T.; Chua, J.K.; Fan, D.S.; Lam, D.S. The association between cigarette smoking and ocular diseases. *Hong Kong Med. J.* **2000**, *6*, 195–202. [PubMed]
27. Sharma, N.; Aggarwal, L.M. Automated medical image segmentation techniques. *J. Med. Phys.* **2010**, *35*, 3–14. [CrossRef]
28. Fuadah, Y.; Setiawan, A.; Mengko, T. Performing high accuracy of the system for cataract detection using statistical texture analysis and K-Nearest Neighbor. In Proceedings of the 2015 International Seminar on Intelligent Technology and Its Applications, Surabaya, Indonesia, 20–21 May 2015; pp. 85–88. [CrossRef]
29. Kauh, C.Y.; Blachley, T.S.; Lichter, P.R.; Lee, P.P.; Stein, J.D. Geographic Variation in the Rate and Timing of Cataract Surgery Among US Communities. *JAMA Ophthalmol.* **2016**, *134*, 267–276. [CrossRef]
30. Tielsch, J.M.; Javitt, J.C.; Coleman, A.; Katz, J.; Sommer, A. The Prevalence of Blindness and Visual Impairment among Nursing Home Residents in Baltimore. *N. Engl. J. Med.* **1995**, *332*, 1205–1209. [CrossRef]
31. Mainous, A.G.; Johnson, S.P.; Saxena, S.K.; Wright, R.U. Inpatient Bariatric Surgery among Eligible Black and White Men and Women in the United States, 1999–2010. *Am. J. Gastroenterol.* **2013**, *108*, 1218–1223. [CrossRef]
32. Smedley, B.D.; Stith, A.; Nelson, A.R. Committee on Understanding and Eliminating Racial and Ethnic Disparities in Health Care BoHSPS, Institute of Medicine (IOM) Staff. Introduction and Literature Review. In *Unequal Treatment: Confronting Racial and Ethnic Disparities in Health Care*; Nelson, A.R., Ed.; National Academies Press: Washington, DC, USA, 2002; p. 29.
33. Collins, K.; Hall, A.; Neuhaus, C.U.S. *Minority Health: A Chartbook*; The Commonwealth Fund: New York, NY, USA, 1999.
34. Heron, S.L.; Stettner, E.; Haley, L.L., Jr. Racial and Ethnic Disparities in the Emergency Department: A Public Health Perspective. *Emerg. Med. Clin. N. Am.* **2006**, *24*, 905–923. [CrossRef]
35. Leng, S.; Bruesewitz, M.; Tao, S.; Rajendran, K.; Halaweish, A.F.; Campeau, N.G.; Fletcher, J.G.; McCollough, C.H. Photon-counting Detector CT: System Design and Clinical Applications of an Emerging Technology. *Radiographics* **2019**, *39*, 729–743. [CrossRef]
36. Lamba, R.; McGahan, J.P.; Corwin, M.T.; Li, C.-S.; Tran, T.; Seibert, J.A.; Boone, J.M. CT Hounsfield Numbers of Soft Tissues on Unenhanced Abdominal CT Scans: Variability Between Two Different Manufacturers' MDCT Scanners. *Am. J. Roentgenol.* **2014**, *203*, 1013–1020. [CrossRef] [PubMed]
37. Kalender, W.A. *Computed Tomography: Fundamentals, System Technology, Image Quality, Applications*, 3rd ed.; Publicis Publishing: Erlangen, Germany, 2011.

*diagnostics*

*Article*

# Incoherent Optical Fluctuation Flowmetry: A New Method for the Assessment of Foot Perfusion in Patients with Diabetes-Related Lower-Extremity Complications

**Polina Glazkova [1,*], Alexey Glazkov [1], Dmitry Kulikov [2,3], Sergei Zagarov [1], Yulia Kovaleva [1], Alina Babenko [4], Yulia Kononova [4], Elena Kitaeva [4], Timur Britvin [1], Natalia Mazur [1], Roman Larkov [1] and Dmitry Rogatkin [1]**

1 Moscow Regional Research and Clinical Institute ("MONIKI"), 129110 Moscow, Russia
2 Medical Faculty, Moscow Region State University, 141014 Mytishchi, Russia
3 N.A. Semashko National Research Institute of Public Health, 105064 Moscow, Russia
4 Almazov National Medical Research Centre, 197341 St. Petersburg, Russia
* Correspondence: polinikul@mail.ru

**Abstract:** (1) Background: To date, there are no studies evaluating the ability of the incoherent optical fluctuation flowmetry (IOFF) method to assess foot tissue perfusion. The aim of this study was to evaluate the correlation between perfusion values measured by IOFF and $TcPO_2$ in patients with diabetes-related lower-extremity complications. (2) Methods: This was an observational, cross-sectional, two-center study. Diabetic patients with peripheral artery disease and/or diabetic foot ulcers were studied (n = 27, examinations were carried out on 54 legs). Perfusion in the foot tissues was assessed using $TcPO_2$ (reference standard for this study) and the IOFF method. (3) Results: High correlation coefficients of all perfusion parameters measured by IOFF with $TcPO_2$ (Rs 0.7 to 0.76) were shown. The study demonstrated that the IOFF method allows, with a sensitivity of 85.7% and a specificity of 90.0%, the identification of patients with a critical decrease in $TcPO_2 < 20$ mmHg. (4) Conclusions: The high correlation of IOFF parameters with $TcPO_2$ and the moderately high sensitivity and specificity in detecting patients with severe ischemia of foot tissues shows the promise of the method for assessing a tissue perfusion in patients with diabetes-related lower-extremity complications.

**Keywords:** microcirculation; peripheral artery disease; diabetes mellitus; critical limb ischemia; laser doppler flowmetry; incoherent optical fluctuation flowmetry (IOFF)

**Citation:** Glazkova, P.; Glazkov, A.; Kulikov, D.; Zagarov, S.; Kovaleva, Y.; Babenko, A.; Kononova, Y.; Kitaeva, E.; Britvin, T.; Mazur, N.; et al. Incoherent Optical Fluctuation Flowmetry: A New Method for the Assessment of Foot Perfusion in Patients with Diabetes-Related Lower-Extremity Complications. *Diagnostics* **2022**, *12*, 2922. https://doi.org/10.3390/diagnostics12122922

Academic Editor: Viktor Dremin

Received: 21 October 2022
Accepted: 18 November 2022
Published: 23 November 2022

**Publisher's Note:** MDPI stays neutral with regard to jurisdictional claims in published maps and institutional affiliations.

**Copyright:** © 2022 by the authors. Licensee MDPI, Basel, Switzerland. This article is an open access article distributed under the terms and conditions of the Creative Commons Attribution (CC BY) license (https://creativecommons.org/licenses/by/4.0/).

## 1. Introduction

Diabetic foot syndrome (DFS) is a complication of diabetes mellitus and is described as a group of symptoms including neuropathy, reduced blood supply and infection leading to tissue breakdown, and morbidity that may be followed by amputation. DFS is one of the most serious complications of diabetes mellitus (DM) [1,2]. Peripheral arterial disease (PAD) often has a more aggressive course in diabetic patients and, in turn, can lead to DFS [3,4]. In the population of people living with DM, PAD is characterized by multilevel atherosclerotic lesions as well as greater involvement of the arteries below the knee [5]

Reliable assessment of foot tissue perfusion in patients with limb ischemia is essential for predicting limb outcomes and choosing the treatment algorithm [6]. Measurements of transcutaneous oxygen tension ($TcPO_2$) are widely applied for indirect assessment of foot tissue perfusion in patients with limb ischemia [7]. In a number of studies, $TcPO_2$ measurement has shown predictive value in the management of patients with limb ischemia and diabetic foot ulcers [6,8,9]. This method was included in the Wound, Ischemia, and Infection of the Foot (WIfI) classification system, which is recommended for assessing the risk of limb amputation and potential benefit from successful revascularization in patients with lower extremity atherosclerotic occlusive disease [10]. Thus, the results of $TcPO_2$ assessment can directly influence the management of patients. The significance of this method ($TcPO_2$) increases in patients with DM, as the ankle–brachial index (ABI) may be falsely elevated due to the medial artery calcification [9,11].

However, in addition to a number of clinical limitations (high variability, a dependence on ambient temperature, the presence of edema, a small area of the analyzed tissue, etc.), $TcPO_2$ measurement is also limited by the cost of equipment, consumables, and duration of the measurement [6,12,13]. Alternative methods of assessing foot tissue perfusion have either not found clinical application (laser Doppler flowmetry, laser Doppler imaging) or also have a number of limitations (skin perfusion pressure, 2D perfusion imaging, fluorescence angiography, etc.) [6,9,14].

Thus, the development of new tools for the assessment of tissue perfusion in patients with atherosclerotic peripheral artery disease and DFS is an extremely important task.

An incoherent optical fluctuation flowmetry (IOFF) method has been developed in the Laboratory of Medical and Physical Research, Moscow Regional Research and Clinical Institute ("MONIKI") [15]. This is the first study to assess the clinical perspectives of the method in healthcare.

This study aimed to evaluate the correlation between perfusion values measured by IOFF and $TcPO_2$ values in patients with diabetes-related lower-extremity complications.

## 2. Materials and Methods

### 2.1. Study Design, Patients, and Data Sources

This was an observational, cross-sectional, two-center study. The study was conducted at 2 centers:

A: Moscow Regional Research and Clinical Institute ("MONIKI").

B: Federal State Budgetary Institution "V.A. Almazov National Medical Research Center" of the Ministry of Health of the Russian Federation.

Patients with DM and PAD of lower extremities and/or diabetic foot ulcers were included. PAD was diagnosed by detecting hemodynamically significant stenoses/occlusions of the main arteries of lower extremities. Hemodynamically significant stenoses were defined as the presence of stenosis of at least 50% of the diameter reduction by ultrasound duplex scanning [16–18].

Exclusion criteria: pregnancy; diagnosed systemic autoimmune diseases; severe heart rhythm disorders (atrial fibrillation, frequent extrasystoles); presence of acute respiratory viral infections, fever of any genesis; exacerbation of concomitant chronic diseases; blood diseases—thrombocytopenia, anemia (hemoglobin less than 90 g/L); skin diseases that prevent the study; stage 5 chronic kidney disease (glomerular filtration rate < 15 mL/min/1.73 m$^2$ according to MDRD or CKD-EPI); taking hormone replacement therapy, oral contraceptives; regular use of steroids, nonsteroidal anti-inflammatory drugs (therapy with antiaggregants was not exclusion criterion).

Macro- and microhemodynamics in the vessels of the lower extremities were assessed in all patients. To verify the diagnosis and assess the macrohemodynamic status in the extremities, all patients underwent ultrasound duplex scanning and ABI measurements.

Perfusion in the foot tissues was assessed both using $TcPO_2$ (as the reference standard for this study) and the IOFF method.

$TcPO_2$ was measured by using the TCM4 (Radiometer, Copenhagen, Denmark). The skin was first wiped with an alcohol solution. The sticky fixation ring was fixed at the dorsum of the foot between the first and second metatarsal heads just proximal to the first and second toes. After a few droplets of an electrolyte solution enhancing contact between the electrode and the skin, the electrode was fixed in the fixation ring (Figure 1a). Registration of $TcPO_2$ was carried out when they stabilized after 15–20 min of local heating. A probe temperature of 44 °C was selected. The two feet were measured in sequence.

### 2.2. Perfusion Measurement Using the IOFF Method

The foot tissue perfusion measurement using the IOFF method was performed using prototypes of a new diagnostic device developed by the joint-stock company "Elatma Instrument-Making Enterprise" (Ryazan, Russia).

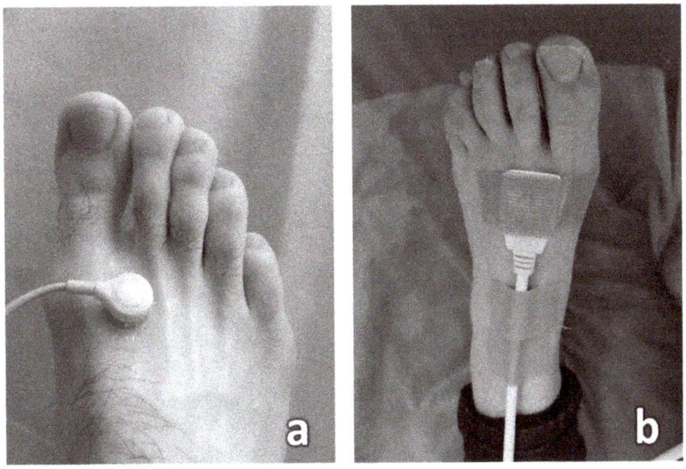


**Figure 1.** (**a**) $TcpO_2$ measurement procedure; (**b**) procedure for measuring perfusion using the IOFF method.

The IOFF method is based on the analysis of low-frequency fluctuations (0–10 Hz) of optical signals backscattered by tissues, initially emitted by the incoherent source—a light-emitting diode (LED). The prototype device uses three LED emission sources operating in the wavelength range of 560–580 nm and one silicon photodiode in the optical sensor. The perfusion value calculated during signal processing is similar to that in laser Doppler flowmetry (LDF) [15,19]. The method allows the assessment of perfusion measured in perfusion units (PU).

The sensor was placed on the dorsal surface of the foot at the first intermetatarsal space (the same location as for the $TcPO_2$ electrode) (Figure 1b). The measurement was carried out sequentially, first on the left leg, then on the right one. A total of 54 measurements were taken in 27 patients.

Perfusion was recorded during a local heating test. The time duration of the test was 6 min. Using heating plates integrated into the perfusion sensor, a thermoneutral temperature of $32 \pm 0.5$ °C was maintained for the first 60 s, and the baseline perfusion level (BP) was assessed. The heating plates were then heated at a rate of 1.5 °C/s to a temperature of $42 \pm 0.5$ °C. This plate temperature was maintained until the end of the measurement (for 5 min). An example of the perfusion curve obtained during the heating test is shown in Figure 2.

The following parameters were assessed:

Baseline perfusion (BP) was calculated as the median baseline perfusion level for the first 60 s of measurement.

Local thermal hyperemia 1–5 min (LTH 1–5 min) was calculated as the median perfusion for each minute of heating.

Before the IOFF and $TcPO_2$ measurements, all patients were at rest for 10 min in the supine position, relaxed, head and heels supported, in a room with a comfortable temperature. All patients were asked to refrain from smoking within 3 h prior to the examinations. Thus, the IOFF perfusion measurement and the $TcPO_2$ measurement were performed at the same location and under the same conditions (patient positioning and preparation). Because the IOFF perfusion measurement and the $TcPO_2$ measurement are accompanied by a heat test, measurements were taken on different days.

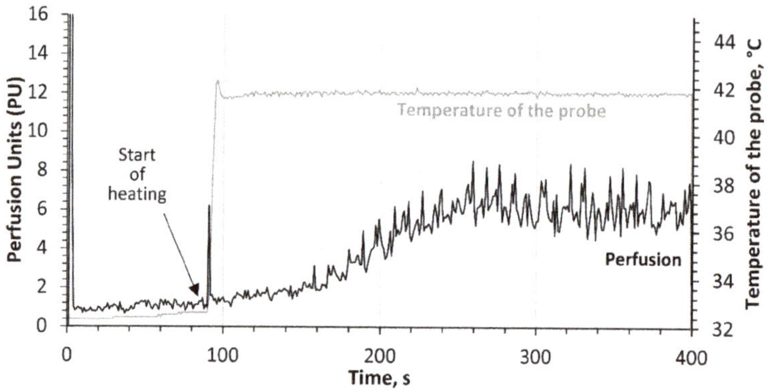


**Figure 2.** Example of a perfusion curve obtained by IOFF during the local heating test.

### *2.3. Assessment of Macrohemodynamic Parameters*

The ABI of each foot was calculated by dividing the higher pressure in the posterior tibial or dorsalis pedis arteries by the higher systolic blood pressure in the right or left arm. To record blood pressure, a Doppler probe was placed over the pulsing artery at a 45° to 60° angle to the surface of the skin [20].

The level and severity of arterial stenosis were assessed by the duplex ultrasound scanning of the lower limb arteries using the Philips Affinity 50 (Philips Ultrasound, Tampa, Florida, USA) and the Vivid 7 Dimension (GE Healthcare, Chicago, Illinois, USA). The ultrasound protocol involved assessing the presence of hemodynamically significant stenoses (at least 50% of the diameter reduction) in 6 main arteries of each lower limb (common femoral artery, deep femoral artery, superficial femoral artery, popliteal artery, anterior tibial artery, posterior tibial artery).

### *2.4. Statistical Analysis*

Statistical analysis of the data was carried out using the RStudio 2021.09.0 Build 351 program using the R version 4.1.1 language. Medians and quartiles (Me (LQ; UQ)) were calculated for quantitative variables. Absolute (n) and relative (%) frequencies were used for qualitative variables. Spearman's correlation coefficient was used to assess the correlation between quantitative variables. Thresholds for quantitative variables were estimated using ROC analysis (pROC 1.18.0 package). The required sample size was calculated using the "power.roc.test()" function. The level of the type I error ($\alpha$) was set equal to 0.05; null hypotheses were rejected at $p < 0.05$.

The required sample size was calculated using the power.roc.test() function from the pROC package. The power of the study was set at 90%. The expected area under the ROC curve for identifying limbs with low $TcPO_2$ (<20 mmHg) was 0.8. The ratio of the number of limbs without $TcPO_2$ reduction to the number of limbs with $TcPO_2$ reduction was set at 2 to 1 ($\kappa = 2$). Thus, at least 39 observations had to be recruited into the study in order to achieve a 90% power level under these conditions. Since the different limbs of the patients were analyzed as independent cases, at least 20 patients had to be included in the study.

## 3. Results

### *3.1. Study Population and Baseline Characteristics*

A total of 27 patients were included in the study. The study was carried out on 54 feet. The characteristics of the group are listed in Table 1.

**Table 1.** Characteristics of the studied patients.

| Parameter | Value |
|---|---|
| Sex: male/female, n (%) | 19 (70.4%)/8 (29.6%) |
| Age, years, Me (LQ; UQ) | 64 (56; 68) |
| Body mass index, kg/m$^2$, Me (LQ; UQ) | 28.4 (26.5; 33.9) |
| Legs with hemodynamically significant artery stenoses by duplex ultrasound | 39 (72.2%) |
| Feet with lower-extremity ulcers, n<br>Neuropathic foot ulcers<br>Ischemic foot ulcers<br>Neuro-ischemic foot ulcers | 14 (51.9%)<br>7 (50%)<br>2 (14.3%)<br>5 (35.7%) |
| ABI (recorded separately for each limb), Me (LQ; UQ) | 0.94 (0.88; 0.98) |
| >1.4, n (%)<br>1–1.4, n (%)<br>0.91–0.99, n (%)<br>0.4–0.9, n (%)<br>0.4, n (%)<br>not defined, n (%) * | 1 (1.8%)<br>7 (13%)<br>18 (33.3%)<br>16 (29.6%)<br>1 (1.8%)<br>11 (20.3%) |
| $TcPO_2$ (recorded separately for each limb), Me (LQ; UQ)<br>$TcPO_2$ < 20 mmHg;<br>20–39 mmHg;<br>$TcPO_2 \geq 40$ mmHg. | 34 (18; 48)<br>14 (25.9%)<br>17 (31.5%)<br>23 (42.6%) |
| MNSI (Part A), Me (LQ; UQ) | 9 (7; 9) |
| MNSI (Part B), Me (LQ; UQ) | 7.00 (5.00; 7.75) |
| HbA1c, %, Me (LQ; UQ) | 8.1 (7.12; 9.45) |
| eGFR according to CKD-EPI (mL/min/1.73 m$^2$), Me (LQ; UQ) | 82 (67; 86) |
| **Comorbidities** | |
| Hypertension, n (%) | 25 (92.6%) |
| Chronic heart failure, n (%) | 10 (37%) |
| Angina pectoris, n (%) | 10 (37%) |

ABI—ankle-brachial index; eGFR using CKD-EPI—Estimated Glomerular Filtration Rate according to Chronic Kidney Disease Epidemiology Collaboration; HbA1—glycated hemoglobin; LQ—lower quartile; Me—median; UQ—upper quartile; MNSI—The Michigan Neuropathy Screening Instrument. * ABI was not detected on 11 limbs due to the absence of a pulse in the arteries of the foot or due to severe pain syndrome.

### 3.2. *Comparison of IOFF and $TcPO_2$ Measurement Results*

It was revealed that all perfusion parameters analyzed by the IOFF method correlated significantly with the $TcPO_2$ measurement with high correlation coefficients (Table 2).

**Table 2.** Correlation between the parameters of the IOFF signal and value of the $TcPO_2$ assessed on 54 measurements (27 patients). The table shows the correlation coefficients.

| | $TcPO_2$ | *p* |
|---|---|---|
| BP | 0.70 | <0.001 |
| LTH, 1 min | 0.74 | <0.001 |
| LTH, 2 min | 0.76 | <0.001 |
| LTH, 3 min | 0.73 | <0.001 |
| LTH, 4 min | 0.75 | <0.001 |
| LTH, 5 min | 0.74 | <0.001 |

BP—baseline perfusion; LTH 1–5 min—local thermal hyperemia for each minute of heating; *p*—statistical significance.

In order to assess this phenomenon in detail, all measurements were divided into three subgroups: subgroup 1—$TcPO_2 < 20$ mmHg (n = 14); subgroup 2—$TcPO_2$ 20–39 mmHg (n = 17); subgroup 3—$TcPO_2 \geq 40$ mmHg (n = 23). These $TcPO_2$ limits have been used because, according to a number of studies, ulcer healing and limb prognosis are generally poor if $TcPO_2$ is <20 mmHg and are generally good if >40 mmHg [7].

The perfusion parameters estimated by the IOFF method were analyzed in each of the subgroups separately and then compared (Figure 3).

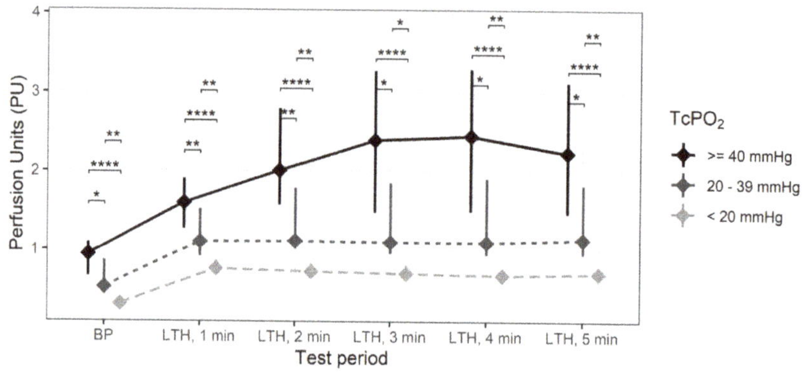


**Figure 3.** Median and interquartile ranges of perfusion measured by IOFF at different time points of the local heating test in legs with different $TcPO_2$ levels (*—$p < 0.05$, **—$p < 0.01$, ****—$p < 0.0001$).

Baseline perfusion and local thermal hyperemia measured by the IOFF method differed significantly in all three subgroups. Limbs with critically low transcutaneous oxygen tension ($TcPO_2 < 20$ mmHg) had a significantly lower baseline perfusion than those in the subgroups 2 and 3. In the first subgroup, the increase in perfusion in response to heating was also significantly less pronounced than in the limbs with higher $TcPO_2$ values. In this subgroup, there was low variability in all analyzed perfusion parameters. In cases where the $TcPO_2$ value was greater than 40 mmHg, the levels of BP and LTH were significantly higher than in the subgroups 1 and 2. The hyperemic response peaked at 3 and 4 min of heating.

Receiver operating characteristic (ROC) curve analyses were performed to assess the ability to identify patients with a critical decrease in $TcPO_2$ (<20 mmHg) based on the IOFF perfusion measurements (Table 3).

**Table 3.** Area under the ROC curve showing the diagnostic potential of IOFF and ABI for identification legs with a critical decrease in $TcPO_2$ (<20 mmHg).

| | AUC | LCL | UCL |
|---|---|---|---|
| ABI | 0.881 | 0.689 | 1 |
| BP | 0.927 | 0.831 | 1 |
| LTH, 1 min | 0.939 | 0.863 | 1 |
| LTH, 2 min | 0.943 | 0.874 | 1 |
| LTH, 3 min | 0.914 | 0.802 | 1 |
| LTH, 4 min | 0.941 | 0.865 | 1 |
| LTH, 5 min | 0.927 | 0.832 | 1 |

ABI—ankle-brachial index; AUC—area under the ROC curve; LCL—lower 95% confidence limit; UCL—upper 95% confidence limit; LTH 1–5 min—local thermal hyperemia for each minute of heating; BP—baseline perfusion.

All analyzed perfusion parameters from the IOFF measurement had a high diagnostic potential in identifying patients with critically low $TcPO_2$. The area under the ROC curve for all IOFF parameters, including BP, was higher than that for ABI.

The area under the curve (AUC) of the LTH, 2 min was 0.943 (95% CI 0.874–1) and had an optimal cutoff value for the identification of limbs with critically low $TcPO_2$ (0.989 PU), with a sensitivity of 85.7% and a specificity of 90% according to the ROC analysis (Figure 4).

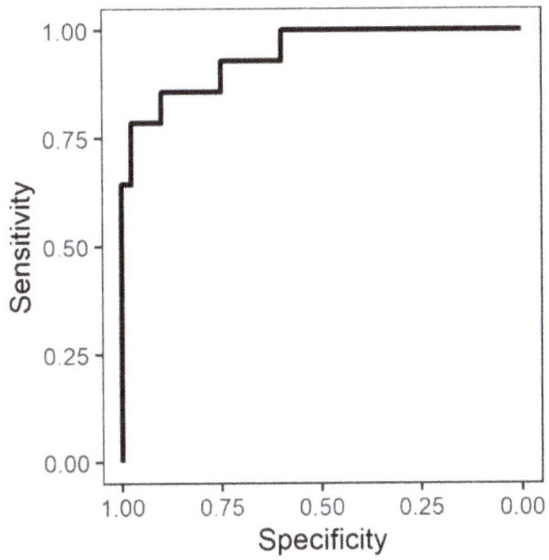


**Figure 4.** ROC curve showing the sensitivity and specificity of LTH, 2 min for identification of legs with a critical decrease in $TcPO_2$ (<20 mmHg).

## 4. Discussion

This study was the first to investigate the informative value of the IOFF technique in assessing a foot tissue perfusion in patients with diabetes-related lower-extremity complications. High correlation coefficients of all perfusion parameters measured by IOFF with $TcPO_2$ (Rs 0.7 to 0.76) were demonstrated. The study showed that the IOFF method allows, with a moderately high sensitivity of 85.7% and a specificity of 90.0%, the identification of patients with a critical decrease in $TcPO_2 < 20$ mmHg. Such a decrease in $TcPO_2$ is known to indicate a poor prognosis for ulcer healing and limb preservation, so identifying patients in this group is of a great clinical importance [7,21].

Current assessment standards in lower extremity artery disease focus on macrovascular function with less emphasis on foot tissue perfusion measurements. However, measurement of foot perfusion is extremely important because macro- and microvascular disorders are not always congruent [22]. A variety of noninvasive diagnostic technologies have been proposed as promising methods for assessing foot tissue perfusion (LDF, LDI, Laser speckle contrast imaging, 2D perfusion imaging, fluorescence angiography, near-infrared spectroscopy, cone-beam computed tomography, etc.) [6,22–26]. However, most of these methods have not yet become widespread in clinical practice [6,25].

The ideal method for the assessment of a foot tissue perfusion should be inexpensive, readily available, and reproducible, and its results should help the clinician to predict wound healing and provide information that influences patient management [6]. The development of such new methods is complicated by the lack of a widely applicable, highly sensitive method that could be called the "gold standard" for studying perfusion and microcirculation in tissues. In this study, $TcPO_2$ measurement was used as a reference method. $TcPO_2$ method has been shown to be highly informative in predicting the ulcer healing and foot amputation, but the method is limited in application due to the cost of equipment, consumables, and duration of the measurements [6,12,13]. Thus, the development of new, widely available methods of assessing foot tissue perfusion can significantly improve the quality of management of patients with diabetes-related lower-extremity complications.

A number of studies have evaluated the applicability of LDF and laser Doppler imaging as a noninvasive optical method for assessing a tissue perfusion [6,27–30]. However, these tools have not found widespread clinical application due to the high frequency of artefacts, sensitivity to ambient temperature, poor reproducibility, and high operator dependence. In addition, the disadvantage of the LDF method is the very small volume of probed tissue (1 $mm^3$ or smaller) [9,31]. Previously, LDF has been shown to be less predictive of ulcer healing and forefoot amputation than the $TcPO_2$ testing [28].

The new IOFF method makes it possible to analyze the perfusion index similar to that calculated by the LDF method. The use of LEDs makes it possible to significantly reduce the cost of the sensors and device. LEDs also eliminate the need for lasers and fiber optics. This reduces the impact of wire positioning on instrument performance. The penetration depth at IOFF is 2–3 mm, which is slightly greater than in LDF. In addition, because of the probe design, the signal is analyzed from a larger area of skin (~25 $mm^2$). Due to this, the signal backscattered from the tissue is collected from a larger tissue volume than in LDF and includes deeper vascular plexuses and larger vessels [14,15]. This also reduces the impact of local vascular network heterogeneity on measurement variability.

The assessment of perfusion by IOFF and $TcPO_2$ are similar. Both procedures involve the local heating test, and the measurement results are affected by a tissue blood supply. However, perfusion assessment using the IOFF method is significantly less time consuming than measuring $TcPO_2$ (6 min vs. 20 min). In addition, the IOFF method does not require expensive consumables, and the use of the light-emitting diodes makes the technology easy to implement and potentially widely available. Expected cost of IOFF devices would be about USD 2000. This low estimated cost is an important potential advantage of the IOFF method over other methods of perfusion assessment (such as LDF, $TcPO_2$, 2D perfusion imaging, cone-beam computed tomography, etc.)

Thus, the new method IOFF may be promising as an informative, available and convenient way to assess foot tissue perfusion in patients with diabetes-related lower-extremity complications.

Additionally in this study, the IOFF method was compared with the results of measuring ABI. The area under the ROC curve for all IOFF parameters was higher than that for ABI for identification legs with a critical ischemia. However, it is known that the usefulness of the ABI is limited in people with DM because of medial arterial calcification [11]. Thus, in this population, the toe–brachial index (TBI) may be more informative [32]. However, this index is not applicable for patients with an amputated big toe. Though TBI was not measured in our study, it will be of great interest to compare the results of TBI and IOFF assessment in further studies.

This first study of the IOFF technique foreshadows further necessary longitudinal research, which should focus on endpoint analysis and the derivation of specific perfusion thresholds for the probability of wound healing. For a more accurate assessment of the applicability of the method IOFF in practical healthcare, further study of reproducibility, testing of the methodology on large sample sizes, and evaluation of the predictive ability of the method is required. It is also of great interest to explore the association between flow and perfusion [33] and to assess the diagnostic informativeness of the method as a screening tool for the detection of PAD in both diabetic and nondiabetic patients.

## 5. Limitations

In this study, we used strict inclusion/exclusion criteria. This reduced possible measurement bias (e.g., associated with heart rhythm disorders or anemia) but may affect the reproducibility of the result in "real-world practice".

It is known that the duplex ultrasound scanning is an operator-dependent procedure. The study was conducted at two independent centers, so the assessment of the duplex ultrasound scanning was performed by several experts, which could affect the accuracy.

## 6. Conclusions

The results of the pilot study demonstrated a high correlation between the perfusion parameters assessed by IOFF and $TcPO_2$. A sensitivity of 85.7% and specificity of 90.0% in identifying patients with critically decreased $TcPO_2 < 20$ mmHg suggests that the IOFF technique may be promising as an informative, rapid, and noninvasive method of assessing tissue perfusion.

**Author Contributions:** Conceptualization, P.G., A.G., D.K. and D.R.; methodology, P.G., A.G., D.K., Y.K. (Yulia Kovaleva), A.B., Y.K. (Yulia Kononova) and E.K.; software, A.G.; validation, A.G., R.L., T.B. and N.M.; formal analysis, A.G.; investigation, P.G., A.G., S.Z., Y.K. (Yulia Kovaleva), A.B., Y.K. (Yulia Kononova), E.K. and N.M.; resources, D.K., S.Z., A.B., T.B., R.L. and D.R.; data curation, P.G., A.G., S.Z., Y.K. (Yulia Kovaleva), A.B., Y.K. (Yulia Kononova) and N.M.; writing—original draft preparation, P.G.; writing—review and editing, A.G., D.K., S.Z., Y.K. (Yulia Kovaleva), A.B., Y.K. (Yulia Kononova), E.K., T.B., N.M., R.L. and D.R.; visualization, A.G.; supervision, D.K., A.B. and D.R.; project administration, D.K., A.B. and D.R.; funding acquisition, D.R. All authors have read and agreed to the published version of the manuscript.

**Funding:** The recruitment and examination of patients was funded by JSC "Elatma Instrument-Making Enterprise" (Ryazan, Russia). Data analysis and evaluation was carried out by researchers without funding from JSC "Elatma Instrument-Making Enterprise" within the framework of the research project "New Approaches to the Comprehensive Assessment of Peripheral Hemodynamic Parameters in the Management of Patients with Diseases of Various Etiologies", funded by the State Budget of the Moscow Region. The funding source had no role in this manuscript.

**Institutional Review Board Statement:** The study was conducted in accordance with the Declaration of Helsinki. The trial protocol was approved by Research Ethics Committees at participating institutions: Moscow Regional Research and Clinical Institute independent ethics committee (Protocol No. 13 dated 7 November 2019) and Almazov National Medical Research Centre ethics committee (No. 27112019, meeting No.11-19 dated 11 November 2019). The photograph of the subjects' legs was accompanied by the written permission of the subjects.

**Informed Consent Statement:** Written informed consent was obtained from all subjects involved in the study.

**Data Availability Statement:** The data that support the findings of this study are available from the corresponding author (Glazkova P.) upon reasonable request.

**Conflicts of Interest:** The authors declared no potential conflict of interest with respect to the research, authorship, and/or publication of this article.

## References

1. Schaper, N.C.; van Netten, J.J.; Apelqvist, J.; Bus, S.A.; Hinchliffe, R.J.; Lipsky, B.A. IWGDF Editorial Board. Practical Guidelines on the prevention and management of diabetic foot disease (IWGDF 2019 update). *Diabetes Metab. Res. Rev.* **2020**, *36*, e3266. [CrossRef] [PubMed]
2. Zou, S.; Zhao, Y.; Shen, Y.; Shi, Y.; Zhou, H.; Zou, J.; Shi, B. Identifying at-risk foot among hospitalized patients with type 2 diabetes: A cross-sectional study in one Chinese tertiary hospital. *Chronic Dis. Transl. Med.* **2015**, *1*, 210–216. [CrossRef]
3. Balletshofer, B.; Böckler, D.; Diener, H.; Heckenkamp, J.; Ito, W.; Katoh, M.; Lawall, H.; Malyar, N.; Oberländer, Y.; Reimer, P.; et al. Position Paper on the Diagnosis and Treatment of Peripheral Arterial Disease (PAD) in People with Diabetes Mellitus. *Exp. Clin. Endocrinol. Diabetes* **2019**, *127*, S105–S113. [CrossRef] [PubMed]
4. Luan, J.; Xu, J.; Zhong, W.; Zhou, Y.; Liu, H.; Qian, K. Adverse prognosis of peripheral artery disease treatments associated with diabetes: A comprehensive meta-analysis. *Angiology* **2022**, *73*, 318–330. [CrossRef] [PubMed]
5. Jakubiak, G.K.; Pawlas, N.; Cieślar, G.; Stanek, A. Pathogenesis and clinical significance of in-stent restenosis in patients with diabetes. *Int. J. Environ. Res. Public Health* **2021**, *18*, 11970. [CrossRef]
6. Rogers, R.K.; Montero-Baker, M.; Biswas, M.; Morrison, J.; Braun, J. Assessment of foot perfusion: Overview of modalities, review of evidence, and identification of evidence gaps. *Vasc. Med.* **2020**, *25*, 235–245. [CrossRef]
7. Leenstra, B.; Wijnand, J.; Verhoeven, B.; Koning, O.; Teraa, M.; Verhaar, M.C.; de Borst, G.J. Applicability of Transcutaneous Oxygen Tension Measurement in the Assessment of Chronic Limb-Threatening Ischemia. *Angiology* **2020**, *71*, 208–216. [CrossRef]
8. Wang, Z.; Hasan, R.; Firwana, B.; Elraiyah, T.; Tsapas, A.; Prokop, L.; Mills, J.L.; Murad, M.H. A systematic review and meta-analysis of tests to predict wound healing in diabetic foot. *J. Vasc. Surg.* **2016**, *63*, 29S–36S. [CrossRef]
9. Forsythe, R.O.; Hinchliffe, R.J. Assessment of foot perfusion in patients with a diabetic foot ulcer. *Diabetes Metab. Res. Rev.* **2016**, *32*, 232–238. [CrossRef]
10. Mills Sr, J.L.; Conte, M.S.; Armstrong, D.G.; Pomposelli, F.B.; Schanzer, A.; Sidawy, A.N.; Andros, G. Society for Vascular Surgery Lower Extremity Guidelines Committee. The society for vascular surgery lower extremity threatened limb classification system: Risk stratification based on Wound, Ischemia, and foot Infection (WIfI). *J. Vasc. Surg.* **2014**, *59*, 220–234. [CrossRef]

11. Potier, L.; Abi Khalil, C.; Mohammedi, K.; Roussel, R. Use and utility of Ankle brachial index in patients with diabetes. *Eur. J. Vasc. Endovasc. Surg.* **2011**, *41*, 110–116. [CrossRef] [PubMed]
12. Costanzo, I.; Sen, D.; Adegite, J.; Rao, P.M.; Guler, U. Noninvasive Miniaturized Transcutaneous Oxygen Monitor. *IEEE Trans. Biomed. Circuits Syst.* **2021**, *15*, 474–485. [CrossRef] [PubMed]
13. Bondarenko, O.N.; Ayubova, N.L.; Galstyan, G.R.; Dedov, I.I. Transcutaneous oximetry monitoring in patients with type 2 diabetes mellitus and critical limb ischemia. *Diabetes Mellit.* **2013**, *16*, 33–42. (In Russian) [CrossRef]
14. Bajwa, A.; Wesolowski, R.; Patel, A.; Saha, P.; Ludwinski, F.; Smith, A.; Nagel, E.; Modarai, B. Assessment of tissue perfusion in the lower limb current methods and techniques under development. *Circ. Cardiovasc. Imaging* **2014**, *7*, 836–843. [CrossRef]
15. Lapitan, D.; Rogatkin, D. Optical incoherent technique for noninvasive assessment of blood flow in tissues: Theoretical model and experimental study. *J. Biophotonics* **2021**, *14*, e202000459. [CrossRef]
16. Siao, R.M.; So, M.J.; Gomez, M.H. Pulse oximetry as a screening test for hemodynamically significant lower extremity peripheral artery disease in adults with type 2 diabetes mellitus. *J. ASEAN Fed. Endocr. Soc.* **2018**, *33*, 130–136. [CrossRef]
17. Varga-Szemes, A.; Penmetsa, M.; Emrich, T.; Todoran, T.M.; Suranyi, P.; Fuller, S.R.; Edelman, R.R.; Kok-tzoglou, I.; Schoepf, U.J. Diagnostic accuracy of non-contrast quiescent-interval slice-selective (QISS) MRA combined with MRI-based vascular calcification visualization for the assessment of arterial stenosis in patients with lower extremity peripheral artery disease. *Eur. Radiol.* **2021**, *31*, 2778–2787. [CrossRef]
18. Armstrong, P.A.; Dargan, C.G. (Eds.) Duplex Scanning for Lower Extremity Arterial Disease. In *Noninvasive Vascular Diagnosis: A Practical Textbook for Clinicians*; Springer International Publishing: Cham, Switzerland, 2022; p. 535.
19. Lapitan, D.G.; Raznitsyn, O.A. A Method and a Device Prototype for Noninvasive Measurements of Blood Perfusion in a Tissue. *Instrum. Exp. Tech.* **2018**, *61*, 745–750. [CrossRef]
20. Aboyans, V.; Criqui, M.H.; Abraham, P.; Allison, M.A.; Creager, M.A.; Diehm, C.; Gerry, F.; Fowkes, R.; Hiatt, W.R.; Jönsson, B.; et al. Measurement and interpretation of the Ankle-Brachial Index: A scientific statement from the American Heart Association. *Circulation* **2012**, *126*, 2890–2909. [CrossRef]
21. Catella, J.; Long, A.; Mazzolai, L. What is currently the role of TcPO2 in the choice of the amputation level of lower limbs? A comprehensive review. *J. Clin. Med.* **2021**, *10*, 1413. [CrossRef]
22. Geskin, G.; Mulock, M.D.; Tomko, N.L.; Dasta, A.; Gopalakrishnan, S. Effects of Lower Limb Revascularization on the Microcirculation of the Foot: A Retrospective Cohort Study. *Diagnostics* **2022**, *12*, 1320. [CrossRef] [PubMed]
23. Filina, M.A.; Potapova, E.V.; Makovik, I.N.; Zharkih, E.V.; Dremin, V.V.; Zherebtsov, E.A.; Dunaev, A.V.; Sidorov, V.V.; Krupatkin, A.I.; Alimicheva, E.A.; et al. Functional changes in blood microcirculation in the skin of the foot during heating tests in patients with diabetes mellitus. *Hum. Physiol.* **2017**, *43*, 693–699. [CrossRef]
24. Dremin, V.V.; Zherebtsov, E.A.; Sidorov, V.V.; Krupatkin, A.I.; Makovik, I.N.; Zherebtsova, A.I.; Zharkikh, E.V.; Potapova, E.V.; Dunaev, A.V.; Doronin, A.A.; et al. Multimodal optical measurement for study of lower limb tissue viability in patients with diabetes mellitus. *J. Biomed. Opt.* **2017**, *22*, 085003. [CrossRef] [PubMed]
25. Zharkikh, E.; Dremin, V.; Zherebtsov, E.; Dunaev, A.; Meglinski, I. Biophotonics methods for functional mo toring of compli-cations of diabetes mellitus. *J. Biophotonics* **2020**, *13*, e202000203. [CrossRef] [PubMed]
26. Kim, R.; Choi, S.Y.; Kim, Y.J. The Efficacy of Cone-Beam CT-Based Perfusion Mapping in Evaluation of Tissue Perfusion in Peripheral Arterial Disease. *J. Clin. Med.* **2021**, *10*, 947. [CrossRef] [PubMed]
27. Lantsberg, L.; Goldman, M. Laser doppler flowmetry, transcutaneous oxygen tension measurements and doppler pressure compared in patients undergoing amputation. *Eur. J. Vasc. Surg.* **1991**, *5*, 195–197. [CrossRef]
28. Karanfilian, R.G.; Lynch, T.G.; Zirul, V.T.; Padberg, F.T.; Jamil, Z.; Hobson II, R.W. The value of laser Doppler velocimetry and transcutaneous oxygen tension determination in predicting healing of ischemic forefoot ulcerations and amputations in diabetic and nondiabetic patients. *J. Vasc. Surg.* **1986**, *4*, 511–516. [CrossRef]
29. Talitskiy, K.A.; Bulkina, O.S.; Aref'eva, T.I.; Vorob'eva, O.N.; Levitskiy, I.V.; Fedorovich, A.A. Effi cacy of therapeutic angiogenesis in patients with chronic lower limb ischemia. *Cell. Transpl. Tissue Eng.* **2011**, *6*, 89–98.
30. Kozlov, I.; Zherebtsov, E.; Masalygina, G.; Podmasteryev, K.; Dunaev, A. Laser Doppler Spectrum Analysis Based on Calculation of Cumulative Sums Detects Changes in Skin Capillary Blood Flow in Type 2 Diabetes Mellitus. *Diagnostics* **2021**, *11*, 267. [CrossRef]
31. Roustit, M.; Cracowski, J.L. Non-invasive assessment of skin microvascular function in humans: An insight into methods. *Microcirculation* **2012**, *19*, 47–64. [CrossRef]
32. Jakubiak, G.K.; Pawlas, N.; Cieślar, G.; Stanek, A. Chronic lower extremity ischemia and its association with the frailty syndrome in patients with diabetes. *Int. J. Environ. Res. Public Health* **2020**, *17*, 9339. [CrossRef] [PubMed]
33. Normahani, P.; Khosravi, S.; Sounderajah, V.; Aslam, M.; Standfield, N.J.; Jaffer, U. The effect of lower limb revascularization on flow, perfusion, and systemic endothelial function: A systematic review. *Angiology* **2021**, *72*, 210–220. [CrossRef] [PubMed]

diagnostics

MDPI

Article

# Biochemical and Mechanical Analysis of Occlusal and Proximal Carious Lesions

**Sahar Al-Shareefi [1], Ali Addie [2] and Lamis Al-Taee [1,*]**

1 Department of Conservative and Esthetic Dentistry, Baghdad College of Dentistry, University of Baghdad, Baghdad P.O. Box 1417, Iraq
2 Center of Advanced Materials, Ministry of Science and Technology, Baghdad P.O. Box 0765, Iraq
* Correspondence: lamis.al-taee@codental.uobaghdad.edu.iq; Tel.: +964-772-812-0182

**Abstract:** A precise evaluation of caries excavation endpoint is essential in clinical and laboratory investigations. Caries invasion differentiates dentin into structurally altered layers. This study assessed these changes using Raman spectroscopy and Vickers microhardness. Ten permanent molars with occlusal and proximal carious lesions were assessed and compared at 130 points utilizing four Raman spectroscopic peaks: phosphate v1 at 960 $cm^{-1}$, amide I (1650 $cm^{-1}$), amide III (1235 $cm^{-1}$) and the C-H bond of the pyrrolidine ring (1450 $cm^{-1}$). The phosphate-to-amide I peak ratio and collagen integrity peak ratio (amide III: C-H bond) of carious zones were calculated and compared in both lesions. The former ratio was correlated to 130 Vickers microhardness indentations through lesions. The caries-infected dentin (CID) exhibited low phosphate peak, but higher amide I, III and C-H bond peaks than other zones in both lesions. The peaks in amide regions (I and III) varied in occlusal versus proximal lesions. A high correlation was found between mineral: matrix peak ratio and equivalent microhardness number within carious lesions, while the collagen integrity peak ratio was applied in proximal lesions only. Raman spectroscopy detected changes in the mineral and matrix contents within different carious zones and regions.

**Keywords:** caries-infected dentin; caries-affected dentin; collagen integrity; Raman microscopy; Vickers microhardness

**Citation:** Al-Shareefi, S.; Addie, A.; Al-Taee, L. Biochemical and Mechanical Analysis of Occlusal and Proximal Carious Lesions. *Diagnostics* **2022**, *12*, 2944. https://doi.org/10.3390/diagnostics12122944

Academic Editors: Daniel Fried and Viktor Dremin

Received: 28 September 2022
Accepted: 19 November 2022
Published: 25 November 2022

**Publisher's Note:** MDPI stays neutral with regard to jurisdictional claims in published maps and institutional affiliations.

**Copyright:** © 2022 by the authors. Licensee MDPI, Basel, Switzerland. This article is an open access article distributed under the terms and conditions of the Creative Commons Attribution (CC BY) license (https://creativecommons.org/licenses/by/4.0/).

## 1. Introduction

Dental caries is the most common multifaceted disease with a significant societal impact that spreads worldwide [1]. The occlusal and proximal tooth surfaces are the most susceptible sites for demineralization from the bacterial acidic by-products [2]. Although the predominance of occlusal caries was higher [3], in which the bacterial accumulations receive the best protection in the deepest part of the groove-fossa system, carious progression in smooth surfaces is also popular due to the difficulty of early detection using the standard visual-tactile evaluation [4]. However, the radiographs and fiber-optic transillumination can be beneficial, but they cannot always lead to a definitive diagnosis [5]. Carious dentin is classified clinically and histologically into two main zones, superficial infected (CID) and deeper affected layer (CAD), in which the delineation between them is critical in clinical and laboratory investigations. The use of minimally invasive caries management reduces unnecessary tissue removal and the risk of pulp exposure, whilst maximizing the reparative potential of the dentin-pulp complex [6]. This includes complete excavation of the superficial, bacterially contaminated and denatured dentin while preserving the remaining harder affected dentin that can be sealed with therapeutic restorations [7]. Clinically, the discrimination between the carious zones based on tissue hardness, moisture, color, fluorescence properties and dye stainability. Additionally, carious tissue removal methods can be used to determine the endpoint, such as self-limiting burs and chemomechanical removal agents. These methods are implemented in in vitro investigations but lack sufficient clinical validation [8]. The laboratory studies assessed the mechanical properties of the carious dentin through relative changes in tissue hardness via Knoop or Vickers hardness test as a gold standard to delineate the excavation margins. However, this test is considered as an invasive, low-resolution test that damages tissues irreversibly preventing further tissue analysis [9]. Furthermore, hardness clarifies the mechanical integrity without any correlation to the biochemistry of the carious lesion. In contrast, the use non-invasive analytical technology for in vitro studies such as Raman spectroscopy will help in assessing the biochemical tissue changes during the caries process. It can be sufficiently sensitive to the differences in the mineral and organic compositions of healthy and carious dentin tissues through their specific molecular vibrational energy signatures [10]. Previous studies utilized FTIR and Raman spectroscopy to identify the spectral characteristics of mineral-to-matrix peak ratio, which represents the phosphate and amide I distributions through the carious dentin lesion [11–14]. The phosphate peak intensity is related to the mineral content, while the amide I peak intensity represents the organic component in dentin [15–18]. Thus, this correlation is regarded as a biochemical measure for dentin demineralization, in which the corresponding hardness values will help to identify the lesion characteristics in each zone. The evaluation of the collagen in carious dentin has a significant importance, since it affects the bond stability when bonded to adhesive resin restorations. The integrity of the collagen triple helix in type I collagen membrane obtained from bovine Achilles deep tendon was previously measured using attenuated total reflection Fourier transform infrared (ATR-FTIR) [19]. In this method, a peak ratio of amide III: C-H bond of the pyrrolidine ring (1235 $cm^{-1}$:1450 $cm^{-1}$) was used as an indicator for the collagen integrity if it is close to 1 [20]. However, this ratio was not tested on a natural dentin carious lesion using Raman microscopy, to validate using this ratio as an indicator for the collagen integrity in vitro investigations. The biochemical characterization of proximal carious lesions is scarce in the literature. The susceptibility of different tooth surfaces for caries development and progression is markedly differed in occlusal versus smooth surfaces [21], added to differences in the inflammatory pulp response in proximal lesions as compared to occlusal lesions [22]. This might necessitate an assessment of the structural changes in proximal carious lesions in comparison to occlusal in the same tooth model. Accordingly, this study evaluated and compared the changes in the mineral content and collagen matrix in different carious zones of occlusal vs. proximal carious lesions using Raman microscopy. The null hypotheses: (i) there were statistically no significant changes in the mineral and matrix contents of demineralized carious lesions (CID and CAD) from sound dentin in each lesion and between lesions, (ii) the collagen integrity ratio (Amide III: CH- bond of the pyrrolidine ring) cannot be applied for a natural carious dentin lesion, and (iii) there was no statistical correlation between Raman peak ratio of phosphate ν1:amide I ratio and the equivalent VHN in occlusal and proximal carious dentin lesions.

## 2. Materials and Methods

### 2.1. Sample Preparation for Raman and Hardness Measurements

Ten permanent, freshly extracted human 1st molars were collected using an ethics protocol approved by the health research committee (Ref No. 285521, 31 March 2021). The teeth showed two active carious dentin lesions; occlusal and proximal at the contact area and above the cementoenamel junction. The lesions showed score 4 following the international caries detection and assessment system (ICDAS) [23], in which the lesion extended halfway through the dentin without pulp exposure, then were stored in distilled water in a cold cabinet (+4 °C). Samples were hemi-sectioned longitudinally (Isomet 1000, Buehler, Lake Bluff, IL, USA) using a water-cooled diamond blade (330-CA/RS-70300, Struers, Detroit Rd. Westlake, LLC, Cleveland, OH, USA), and then embedded in epoxy resin molds. The surfaces were polished under running water using a polishing machine (Laryee Technology, China), and silicon waterproof papers in a sequential pattern (P1200 for 10 s, P2500 for 10 s, and P4000 for 4 min) [13] to gain flat and smooth surfaces for accurate measurements. A small steel round rotary bur was used to create a reference dot at the enamel–dentin junction occlusally and proximally. These dots are visible under Raman spectroscopy and Vickers microhardness with a 500 µm distance between the examined points, as shown in Figure 1.

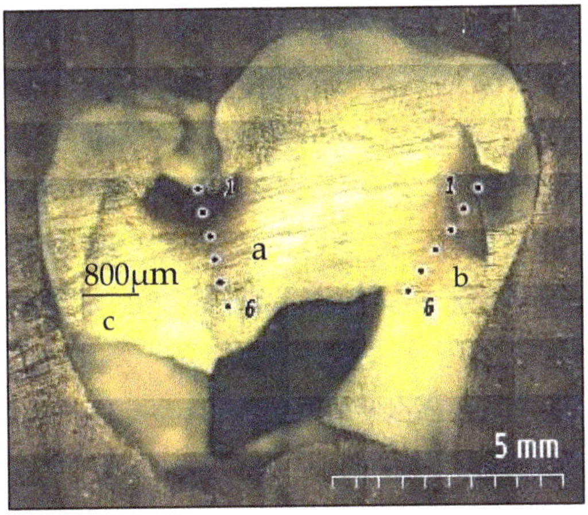


**Figure 1.** The prepared lesion hemi-section was firstly analyzed using confocal Raman microscopy followed by Vickers microhardness indenter across the lesion. The assessment started from the enamel–dentin junction through the lesions towards the pulp in a straight path occlusally (a) and an oblique path proximally (b). A control measurement was taken in an area of clinically sound dentin (c) in the same sample. Six measurement areas were taken in each lesion with a 500 µm distance between the examined points.

### 2.2. *Raman Spectroscopy*

A high-resolution confocal Raman microscope (Senterra, Bruker Optics, Ettlingen, Germany) operating in line scan mode was used to scan the carious lesions and sound dentin. A total of 130-point scans were made over twenty carious lesions ($n$ = 13 per tooth). The distance between the scanned points was 500 µm, and it was controlled using a programmable sample stage with a 1µm resolution. Spectra acquisition was performed using a 785 nm near-infrared diode laser and a 400 line/mm diffraction grating. An Olympus 20X/0.40 NA objective lens was used to focus the laser on the sample surface with a spot size of about 5 µm. The sectioned tooth was mounted on the sample stage using plastic molding putty, and the Raman spectra for carious lesions and sound dentin were measured over the range of 200–3600 cm$^{-1}$ with 100 mW of laser power on each point. The integration time for each spectrum was typically 30 s, with three accumulations. Spectrum acquisition was conducted using the image stitching technique to collect a large area of the mounted tooth (Figure 1). Baseline correction was performed by Raman processing software (OPUS, Bruker Optics, Germany). After acquisition and spectra processing, four Raman spectroscopic peaks were identified. The phosphate peak intensity v1 vibration at 960 cm$^{-1}$, amide I peak intensity at 1650 cm$^{-1}$, amide III at 1235 cm$^{-1}$ and C-H bond of the pyrrolidine ring at 1450 cm$^{-1}$. Peak height intensities were calculated and averaged in both lesions (occlusal and proximal) and the control point (sound dentin). The inorganic to organic ratios of dentin components were assessed from the band intensities of phosphate V1 at (960 cm$^{-1}$) to amide I (1650 cm$^{-1}$). While the collagen integrity at each zone was assessed by calculating the absorbance ratio of 1235 cm$^{-1}$:1450 cm$^{-1}$ ($n$ = 2 per zone, then averaged).

### 2.3. *Vickers Microhardness*

The microhardness of demineralized dentin zones (caries-infected, caries-affected) and sound dentin was measured using the Vickers microhardness tester (TH714, Obsnap Instruments Sdn Bhd, Selangor, Malaysia). A square-based pyramid diamond-shaped indenter was used, with a load of 300 gf for 15 s [24]. A total of 130 indentations ($n$ = 13 per lesion) were made in the same straight path line that previously assessed by Raman microscopy. These indentations started from the enamel–dentin junction occlusally and proximally toward the pulp, with 500 μm interval between the examined points (Figure 1). Then, the Vickers hardness number was recorded automatically using the manufacturer's software.

*2.4. Statistical Analysis and Spectral Correlation Processing with Hardness Measurements*

The statistical analysis was performed using SPSS software version 25 (IBM, Chicago, IL, USA). Shapiro–Wilk test was used to evaluate the normality of data distribution. The data were statistically analyzed using one-way ANOVA followed by Tukey post hoc multiple comparisons ($p > 0.05$) regarding the intensities of four Raman peaks (A.U.) and Vickers microhardness number (VHN) in each zone per lesion. The independent t-test (Minitab 14, Minitab LLC, Chicago, IL, USA) was performed for pairwise comparisons between groups ($\alpha = 0.05$) to assess the differences in these data between the occlusal and proximal lesions at each zone. Additionally, Pearson's correlation coefficient test was used to explore if there is a correlation between Raman mineral: matrix peak ratio and their equivalent microhardness values (VHN) at each point per lesion. The peak ratios were determined after baseline correction in single spectra by dividing the intensity of phosphate to the amide I (960 cm$^{-1}$/1650 cm$^{-1}$).

## 3. Results

*3.1. Biochemical Analysis of Sound and Demineralized Zones in Occlusal and Proximal Lesions*

The relative Raman band intensities (Mean ± SD) with statistical correlations in caries-infected, caries-affected and sound dentin zones of the occlusal and proximal carious lesions are shown in Table 1. All dentin layers (sound and demineralized) showed the evidence of the four characteristic peaks; the phosphate peak intensity (symmetric P-O stretching mode, v1-PO) at 960 cm$^{-1}$ which represents the inorganic part of dentin, the amide I peak intensity at 1650 cm$^{-1}$, amide III at 1235 cm$^{-1}$ and the C-H bond of the pyrrolidine ring at 1450 cm$^{-1}$ which refer to the organic part of sound and demineralized dentin. One-way ANOVA showed a statistically significant difference between sound and demineralized (lesion) dentin regarding all Raman peaks in both lesions ($p = 0.000$). Fur-ther analysis using Tukey post hoc multiple comparisons revealed that the phosphate band at 960 cm$^{-1}$ in both lesions was statistically the lowest in caries-infected dentin zone (CID, $p = 0.000$), but it was increased with increasing the mineral content in the caries-affected (CAD), and then sound dentin with non-significant difference between occlusal and proximal lesions in each zone (independent $t$-test, $p > 0.05$), Table 1.

In both carious lesions, the intensities of amide I, III and C-H bond of pyrrolidine ring peaks (1650, 1235 and 1450 cm$^{-1}$, respectively) were statistically higher in CID, with no significant difference between CAD and sound dentin ($p > 0.05$), except the C-H bond of pyrrolidine ring peak in the proximal carious lesions which was higher in CAD than in sound dentin ($p = 0.023$), Table 1, Figures 2 and 3.

By comparing the peak intensity between the occlusal and proximal carious lesions, it was found that there were statistically no significant differences in the intensity of the phosphate peak (960 cm$^{-1}$) between both lesions ($p > 0.05$) at each zone. For the Amide I (1650 cm$^{-1}$), the peak intensity in the caries-infected zone was higher in the occlusal lesion as compared to the proximal ($p = 0.014$), while there was statistically no significant difference in CAD and sound dentin zones between lesions ($p > 0.05$). The intensity of amide III peak (1235 cm$^{-1}$) was significantly higher in the occlusal carious lesion in all zones (CID, CAD and sound dentin) in comparison to proximal carious lesion ($p < 0.05$). However, the intensity of C-H bond of the pyrrolidine ring peak at 1450 cm$^{-1}$ was comparable between the occlusal and proximal lesions at each zone ($p > 0.05$).

**Table 1.** Raman band intensities A.U. (Mean ± SD) of caries-infected, caries-affected and sound dentin in occlusal and proximal carious lesions.

| Raman Peaks | Carious Zones ($n = 10$ per Zone) | Mean of Peaks Intensity (Occlusal Lesion) (Mean ± SD) | Mean of Peak Intensity (Proximal Lesion) (Mean ± SD) |
|---|---|---|---|
| Phosphate peak v1-PO (960 $cm^{-1}$) | CID | $89.0 \pm 11.3$ [a] | $79.6 \pm 7.7$ [a] |
| | CAD | $344.4 \pm 95.1$ [b] | $369.6 \pm 84.6$ [b] |
| | Sound | $2105.3 \pm 112.1$ [*c] | $2156.4 \pm 122.7$ [*c] |
| Amide I (1650 $cm^{-1}$) | CID | $84.0 \pm 7.1$ [*] | $68.5 \pm 7.2$ [*^] |
| | CAD | $62.0 \pm 9.6$ [d] | $60.5 \pm 10.9$ [d] |
| | Sound | $60.6 \pm 9.9$ [e] | $51.81 \pm 10.2$ [e] |
| Amide III (1235 $cm^{-1}$) | CID | $112.1 \pm 5.3$ [*] | $69.8 \pm 3.7$ [*^] |
| | CAD | $80.4 \pm 5.7$ | $60.4 \pm 6.0$ [^] |
| | Sound | $78.5 \pm 11.5$ | $56.7 \pm 7.9$ [^] |
| C-H bond of pyrrolidine ring (1450 $cm^{-1}$) | CID | $88.9 \pm 4.9$ [*f] | $85.3 \pm 3.9$ [*f] |
| | CAD | $74.9 \pm 5.5$ [g] | $80.0 \pm 5.7$ [g] |
| | Sound | $70.7 \pm 8.6$ [h] | $68.1 \pm 9.8$ [h] |

(*) significant difference between sound and demineralized dentin (caries-infected or caries-affected), one-way ANOVA test and Tukey post hoc tests (alpha level of 0.05). (^) significant difference between occlusal and proximal carious lesion (Independent *t*-test, $p < 0.05$). Similar letters in rows indicate no significant differences between lesions ($p > 0.05$).

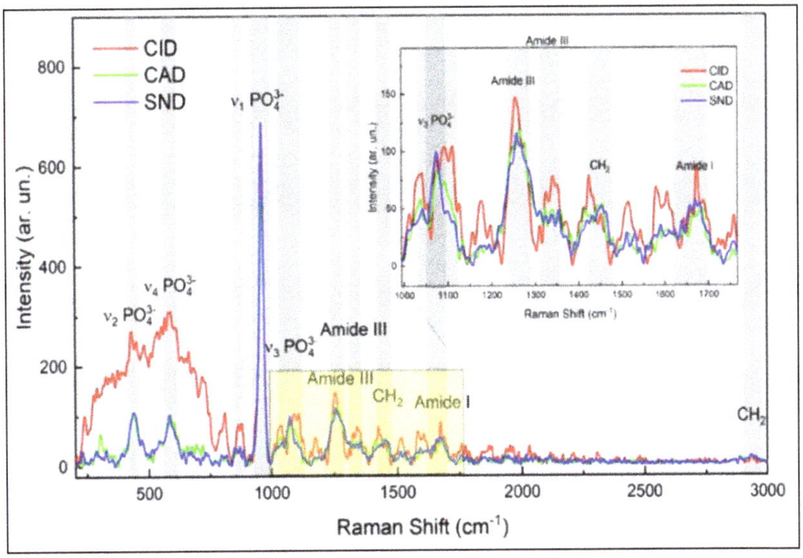


**Figure 2.** Representative Raman spectra of the caries-infected (CID), caries-affected (CAD) and sound dentin in the occlusal carious lesion. The spectra were normalized based on the peak intensities of v1-PO at 960 $cm^{-1}$, which is the strongest signal among all Raman spectra, it is the highest in sound dentin. Amide I at 1650 $cm^{-1}$, Amide III at 1235 $cm^{-1}$ and C-H bond of the pyrrolidine ring at 1450 $cm^{-1}$ are the highest in CID. The inserted figure shows the magnified, same spectra in the range from 1000 to 1700 $cm^{-1}$.

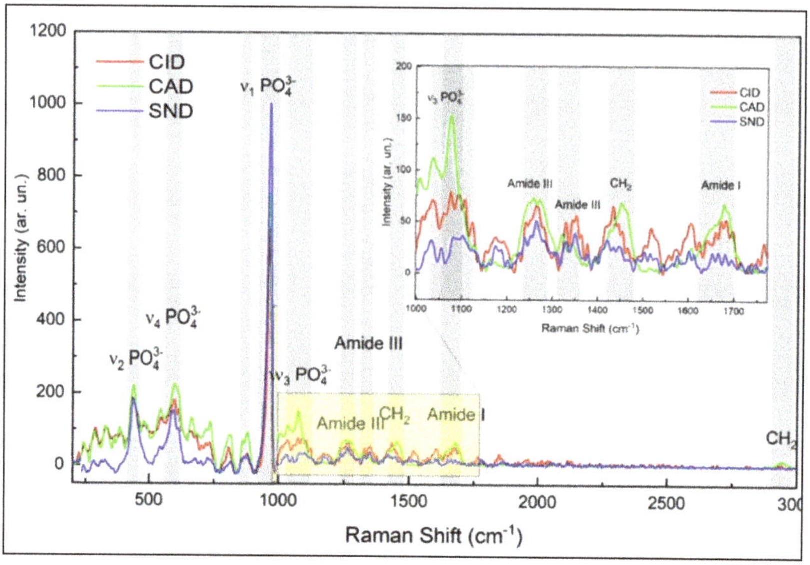


**Figure 3.** Representative Raman spectra of the caries-infected (CID), caries-affected (CAD) and sound dentin in the proximal carious lesion. The spectra were normalized based on the peak intensities of v1-PO at 960 cm$^{-1}$, which is the strongest signal among all Raman spectra, it is the highest in sound dentin. Amide I at 1650 cm$^{-1}$, Amide III at 1235 cm$^{-1}$ and C-H bond of the pyrrolidine ring at 1450 cm$^{-1}$ are higher in CID and CAD than sound dentin. The inserted figure shows the magnified, same spectra in the range from 1000 to 1700 cm$^{-1}$.

To analyze the integrity of the collagen triple helix, a mean absorbance ratio of amide III bands at 1235 cm$^{-1}$ and the pyrrolidine ring at 1450 cm$^{-1}$ was calculated. In the proximal carious lesion, the mean ratio in the caries-infected dentin was 0.7 ± 0.2, while the mean ratio in CAD was 0.9 ± 0.2 that was close to sound dentin (1.0 ± 0.3). This demonstrates a lack of collagen denaturation in the CAD zone. However, this ratio was not applied in the occlusal carious lesion that showed a very high absorbance band of the amide III bands at 1235 cm$^{-1}$ (amide III) than that of the pyrrolidine ring at 1450 cm$^{-1}$, Figure 2.

### 3.2. *Vickers Microhardness*

The change in the mineral content of the carious lesions was also measured via a mi-crohardness test. One-way ANOVA revealed a statistically significant difference among different dentin layers (sound vs. demineralized, $p = 0.000$) with no significant difference found between occlusal and proximal lesions at each zone (independent $t$-test, $p > 0.05$). Further analysis by Tukey post hoc multiple comparisons test showed that there was a marked reduction in VHN in CID at occlusal and proximal lesions (20.2, 22.6, respectively), which was significantly higher in CAD (34.4, 32.5, respectively, $p = 0.000$), followed by sound dentin (54.1, 53.2, respectively, $p = 0.000$), as shown in Figure 4.

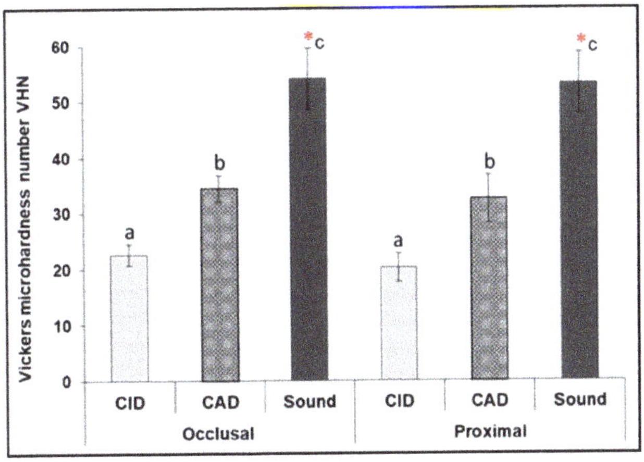


**Figure 4.** Means of Vickers microhardness number (VHN) of the sound and demineralized dentin (caries-infected and caries-affected) in occlusal and proximal lesions. (*) Statistically significant difference of demineralized dentin from sound ($p < 0.001$). Similar litters mean statistically non-significant differences ($p > 0.05$) between different dentin zones in occlusal vs. proximal lesions.

### 3.3. Raman Spectral Correlation with Vickers Microhardness (VHN)

The peak ratios of phosphate v1: amide I of the selected points in each carious lesion (occlusal and proximal) were calculated and plotted against Vickers hardness numbers (VHN) as a scatter diagram, as shown in Figure 5. The coefficient of determination was calculated across the assessed points in the occlusal and proximal lesions, in which the $R^2 = 0.90$ and 0.94, respectively, ($p = 0.000$). The statistically significant high correlation between peak ratio and VHN in both lesions indicates that they showed a logarithmic regression relationship which enable the calculation of the tissue hardness when the peak ratio was measured.

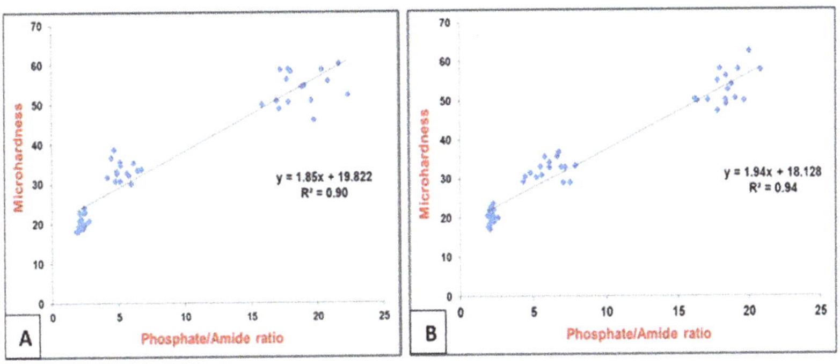


**Figure 5.** A scatter plot and a regression line (R) demonstrating the presence of a logarithmic regression relationship ($p < 0.000$) between the Raman peak ratios (phosphate v1: amide I) and Vickers microhardness (VHN) in the sound and demineralized dentin (CID and CAD) in the occlusal lesion (**A**) and proximal lesion (**B**). The peak ratio is lower in CID with reduced mineral contents which gradually increased towards sound dentin (more phosphate v1).

## 4. Discussion

The improved understanding of the caries process and biology of the dentin-pulp defense and the regenerative responses encouraged the application of minimally invasive caries removal rather than the traditional surgical excavation approach. This approach relies on accurate caries diagnosis, then identifying the excavation endpoint to exclude the irreversibly caries-infected dentin while preserving the remineralizable caries-affected dentin to enhance the long-term survival of the dentin-pulp complexes. Caries invasion leads to the differentiation of dentin into zones with altered composition, collagen integrity and mineral identity. However, the understanding of these changes from the fundamental perspective of molecular structure is limited. Accordingly, this study provided a map of the biochemical changes through the different carious dentin zones in two lesion models (occlusal and proximal) utilizing Raman spectroscopy to extract the molecular information of each zone regarding the hydroxyapatite's structural changes and collagen denaturation as the dentin transition from the superficial caries-infected zone (CID) into sound dentin. The integrity of collagen's triple helical structure was also evaluated based on spectra collected from demineralized dentin (carious lesions) of the selected teeth. The results support the argument that there are statistically significant changes ($p < 0.000$) in the biochemical components across the carious lesions from the superficial layers towards sound dentin.

From a biochemical perspective, the mineral content was detected via the phosphate peak ($PO_4^{-3}$ $\nu 1$) at 960 cm$^{-1}$, which was the strongest signal among all Raman spectra. This peak is referred to the degree of demineralization in natural carious enamel and dentin [25,26]. In accordance with Almahdy et al. (2012) and El-Sharkawy (2019) [18,26], the mineral content was dropped significantly ($p = 0.000$) in CID, but higher in CAD towards sound within the same sample with no significant differences in the intensities between the occlusal and proximal carious lesions ($p > 0.05$) at each zone. The presence of amorphous Ca/P provides a local ion-rich environment, which is favorable for in situ generation of prenucleation clusters, succeeding further dentin remineralization [27]. Accordingly, the first stated hypothesis was partially rejected as the mineral content represented by the phosphate peak intensity differed among different dentin zones (sound and demineralized) but was comparable between occlusal and proximal lesions.

Amide I peak (1650 cm$^{-1}$) is the most prominent organic component of dentin, predominantly type I collagen [28]. This band is assigned to Y8a tyrosine side chain of solution-phase collagen representing the secondary structure of proteins [29]. It was used in many studies to detect changes in the molecular structure of collagen [16,30]. The amide I content was higher in CID in both lesions, and decreased in CAD and sound dentin, as shown in Figures 2 and 3. It was also higher in occlusal than proximal carious lesion. This might be attributed to the higher proteins in the infected layer than sound dentin [31]. The intensity of amide I is correlated to the non-reducible cross-links in collagen [32], which means that there is an apparent change in the molecular structure of collagen in the superficial CID layer in both lesions. This alteration in collagen is correlated to the presence of esters in the carious tissues [17] derived from the bacterial lipid components. This will promote esterification of the carboxylic side-chains of aspartate and glutamate residues catalyzed by the acidic environment (lactic acid) [33]. The closer proximity in the intensity of this band between CAD and sound dentin in both lesions might indicate the importance of preserving this layer since the organic matrix in dentin regulates the growth and maturation of apatite crystals and thus the mineralization process.

The organic component of carious dentin has a major role in the progression of carious development [16,34]. It represents the integral component of the mineralized tissue, but after demineralization, they become exposed and altered structurally [17]. The demineralization process in carious dentin can be assessed by the difference in the mineral/protein band ratio of sound and carious dentin. The band ratio of amide I at ~1650 cm$^{-1}$, which is the most prominent organic moiety in dentin, and the phosphate ion at ~960 cm$^{-1}$ was used in spectroscopic studies [13,14,17,35] to analyze the distribution of mineral content in the dental tissues. Using this band ratio, the Raman peaks graphs showed the heterogeneous nature of caries dentin in terms of mineral distribution with lower values for carious-infected dentin compared to CAD and sound dentin. This band ratio increased by increasing the mineral content in the caries-affected zone followed by sound dentin in both lesions, as shown in Figure 5.

The integrity of the collagen triple helix was evaluated by assessing the ratio of the absorbance bands of amide III at ∼1235 $cm^{-1}$ and CH2 scissoring at ∼1450 $cm^{-1}$ (stereochemistry of the pyrrolidine rings). The former peak is sensitive to the presence of secondary structure of collagen, while the latter is independent of the ordered structure of collagen [20]. The ratio indicates collagen with an integral triple helical structure. If the value is close to 1, but if the ratio <0.8, it means that there is a breakdown of the triple helix in this zone, such as gelatin [19,20]. The mean ratio of amide III: CH2 (1235 $cm^{-1}$:1450 $cm^{-1}$) was 0.7 in CID in the proximal lesion, 0.9 in CAD that was close to sound dentin (1.0) which indicates the integrity of collagen fibers that is essential for the remineralization processes. This supported that the infected layer with comprised denatured collagen, loses the potential for remineralization and must be removed. Conversely, the affected layer, that is partially demineralized and remineralizable with collagen fibrils retaining their natural structure around intact dentinal tubules, is to be preserved to maximize the reparative potential and reduce the risk of pulp exposure [36]. However, this ratio was not achieved in the occlusal carious lesion model. This is due to the higher absorbance of amide III than CH2-CH- bands in the CID zone, as shown in Figure 2, with statistically no significant difference between CAD and sound dentin ($p > 0.05$). This is referred to as the increased aliphatic side-groups of various amino acid residues in the infected tissues in these lesions which promote the demineralization process [37]. The difference between the occlusal and proximal lesions indicate that this ratio cannot be applied for all carious studies due to the variation in collagen denaturation in various tissues and regions. This is due to the variety in depth, degree of organic matrix destruction and loss of mineral content which affect tissue response to different therapeutic materials in an attempt to repair and remineralize the damaged tissues. Accordingly, the second stated hypothesis was partially accepted.

The caries-infected dentin showed higher accentuated amide I and III peaks in occlusal lesions in comparison to proximal carious lesions, which might be due to the higher aliphatic content. This is also might be attributed to the fact that the dentin subjacent to proximal enamel caries forms mineral crystals obliterating the main transport pathways (dentinal tubules), creating sclerotic dentin close to the enamel–dentin junction even before the demineralization reaches the enamel–dentin junction [38]. However, a histological study [39] reported a higher frequency of deep dentin demineralization (>50%) in proximal lesions than occlusal lesions under microradiography with contrast (MRC). This might suggest the presence of facilitated transport pathways for dentinal fluid in dentin underlying the proximal enamel lesion, since the tubules in sclerotic dentin are partially obliterated [40]. This induces a facilitated transport of dentinal fluid into the pores of carious enamel mixed with enamel fluid, and then with bacterial biofilm fluid, and thus promoting the carious progression in proximal lesions.

The microhardness values remain the gold standard for characterizing sound and demineralized dentin tissues in most in vitro studies. It is correlated to the clinical mechanical excavation procedures, and also can predict the behavior of dentin/restoration interfaces, as the regional differences in tissue hardness can alter the distribution of stresses along the interface and thus affecting the preferential location of failures. Additionally, it represents a direct measure of the hydroxyapatite (HAp) crystals present in dental hard tissues. This study showed a gradual increase in tissue hardness from the superficial carious layer towards the deeper area of the carious lesions towards sound dentin ($p = 0.000$). The caries-infected dentin showed the lowest VHN values in both lesions <25, Figure 4, which is attributed to the greater dissolution of HAp compared to the deeper CAD layer (<40) that has the potential to be repaired when sealed with self-adhesive restorations. However, the carious dentin lesion in vivo progresses in a continuous wave rather than distinct zones with gradual transitions in histology and bacteriology from enamel to pulp, it is perhaps rather simplistic to consider the lesion as three distinct zones as described above, but clinically, this analysis has merit because it allows for more reliable, efficient and reproducible operative treatment [41].

Raman can potentially evaluate the carious dentin non-destructively, as an alternative to the invasive, low-resolution hardness tests. This study found statistically significant correlations ($R^2$ = 90, 94%; $p$ = 0.000) between Raman peak ratio (phosphate v1: amide I) and Vickers microhardness number (VHN) in both lesions as shown in Figure 5. This Raman ratio was selected since the mineral content is considered as an indicator of the inherent physical properties of hard tissues. This correlation supports the use of this non-invasive high-resolution technique for chemical characterization in in vitro hard tissue studies rather than the physical tissue microhardness. This has important implications for the tissue pathogenesis and the minimally invasive treatment modalities. Consequently, the third stated hypothesis was rejected.

## 5. Conclusions

Based on the overall results of this study, the use of Raman spectroscopy and subsequent spectra analyses are highly useful for probing the molecular structure of carious dentin in various zones and regions. Raman detected structural changes in the inorganic and organic components of demineralized dentin in comparison to sound dentin in various regions. In which the mineral distribution was the lowest in the caries-infected dentin zone with no differences between occlusal and proximal lesions, whilst the collagen cross-link and triple helical structure were altered differently in carious zones and lesions. The presence of a statistical correlation between the Raman phosphate to amide I peak ratio with VHN supports the use of this non-invasive high-resolution technique for chemical characterization in in vitro studies. This will help further understanding of carious progression and assessing the remaining dentin tissues after different caries removal techniques following the minimally invasive approaches.

**Author Contributions:** S.A.-S.: methodology, investigation, data curation, writing—original draft preparation. A.A.: methodology, resources, analysis. L.A.-T.: conceptualization, methodology, writing—review and editing, supervision. All authors have read and agreed to the published version of the manuscript.

**Funding:** This research received no external funding.

**Institutional Review Board Statement:** Teeth were collected using an ethics protocol approved by health research committee of Baghdad College of Dentistry (Ref No. 285521, 31 March 2021).

**Informed Consent Statement:** Informed consent was obtained from all subjects involved in the study.

**Data Availability Statement:** Not applicable.

**Acknowledgments:** This work was supported by Baghdad College of Dentistry, University of Baghdad, Ministry of Higher Education and Scientific Research/Iraq.

**Conflicts of Interest:** The authors declare that they have no known competing financial interest or personal relationships that could have appeared to influence the study reported in this paper.

## References


1. Paula, J.S.; Leite, I.C.; Almeida, A.B.; Ambrosano, G.; Pereira, A.C.; Mialhe, F.L. The influence of oral health conditions, socioeconomic status and home environment factors on schoolchildren's self-perception of quality of life. *Health Qual. Life Out.* **2012**, *10*, 1–8. [CrossRef] [PubMed]
2. Kidd, E.A.; Fejerskov, O. What constitutes dental caries? Histopathology of carious enamel and dentin related to the action of cariogenic biofilms. *J. Dent. Res.* **2004**, *83* (Suppl. S1), 35–38. [CrossRef]
3. Li, S.H.; Kingman, A.; Forthofer, R.; Swango, P. Comparison of tooth surface-specific dental caries attack patterns in US schoolchildren from two national surveys. *J. Dent. Res.* **1993**, *72*, 1398–1405. [CrossRef]
4. Nakagawa, H.; Sadr, A.; Shimada, Y.; Tagami, J.; Sumi, Y. Validation of swept source optical coherence tomography (SS-OCT) for the diagnosis of smooth surface caries in vitro. *J. Dent.* **2013**, *41*, 80–89. [CrossRef]
5. Baelum, V.; Hintze, H.; Wenzel, A.; Danielsen, B.; Nyvad, B. Implications of caries diagnostic strategies for clinical management decisions. *Community Dent. Oral Epidemiol.* **2012**, *40*, 257–266. [CrossRef] [PubMed]
6. Schwendicke, F.; Frencken, J.; Innes, N. *Caries Excavation, Evolution of Treating Cavitated Carious Lesions*; Karger Medical and Scientific Publishers: Basel, Switzerland, 2018.

7. Banerjee, A.; Pickard, H.M.; Watson, T.F. *Pickard's Manual of Operative Dentistry*, 9th ed.; Oxford University Press: Oxford, UK, 2011.
8. Schwendicke, F.; Dörfer, C.; Paris, S. Incomplete caries removal, a systematic review and meta-analysis. *J. Dent. Res.* **2013**, *92*, 306–314. [CrossRef] [PubMed]
9. Banerjee, A.; Cook, R.; Kellow, S.; Shah, K.; Festy, F.; Sherriff, M.; Watson, T.F. Confocal micro-endoscopic investigation of the relationship between the microhardness of carious dentine and its autofluorescence. *Eur. J. Oral Sci.* **2010**, *118*, 75–79. [CrossRef]
10. Wachsmann-Hogiu, S.; Weeks, T.; Huser, T. Chemical analysis in vivo and in vitro by Raman spectroscopy-from single cells to humans. *Curr. Opin. Biotechnol.* **2009**, *20*, 63–73. [CrossRef]
11. Liu, Y.; Yao, X.; Liu, Y.W.; Wang, Y. A Fourier transform infrared spectroscopy analysis of carious dentin from transparent zone to normal zone. *Caries Res.* **2014**, *48*, 320–329. [CrossRef]
12. Toledano, M.; Osorio, E.; Aguilera, F.S.; Cabello, I.; Toledano-Osorio, M.; Osorio, R. Ex vivo detection and characterization of remineralised carious dentin, by nanoindentation and single point Raman spectroscopy, after amalgam restoration. *J. Raman Spectrosc.* **2017**, *48*, 384–392. [CrossRef]
13. Alturki, M.; Koller, G.; Almhöjd, U.; Banerjee, A. Chemo-mechanical characterization of carious dentine using Raman microscopy and Knoop microhardness. *R. Soc. Open Sci.* **2020**, *7*, 200404. [CrossRef]
14. Alturki, M.; Koller, G.; Warburton, F.; Almhöjd, U.; Banerjee, A. Biochemical characterisation of carious dentine zones using Raman spectroscopy. *J. Dent.* **2021**, *105*, 103558. [CrossRef] [PubMed]
15. Wang, Y.; Spencer, P.; Walker, M.P. Chemical profile of adhesive/caries-affected dentin interfaces using Raman microspectroscopy. *J. Biomed. Mater. Res. A* **2007**, *81*, 279–286. [CrossRef] [PubMed]
16. Lopes, C.D.; Limirio, P.H.; Novais, V.R.; Dechichi, P. Fourier transform infrared spectroscopy (FTIR) application chemical characterization of enamel, dentin and bone. *Appl. Spectrosc. Rev.* **2018**, *53*, 747–769. [CrossRef]
17. Almhöjd, U.S.; Norén, J.G.; Arvidsson, A.; Nilsson, A.; Lingström, P. Analysis of carious dentine using FTIR and ToF-SIMS. *Oral Health Dent Manag.* **2014**, *13*, 735–744. [PubMed]
18. El-Sharkawyi, Y.H. Detection and characterization of human teeth caries using 2D correlation raman spectroscopy. *J. Biomed. Phys. Eng.* **2019**, *9*, 167. [CrossRef]
19. Júnior, Z.S.S.; Botta, S.B.; Ana, P.A.; França, C.M.; Fernandes, K.P.S.; Mesquita-Ferrari, R.A.; Deana, A.; Bussadori, S.K. Effect of papain-based gel on type I collagen-spectroscopy applied for microstructural analysis. *Sci. Rep.* **2015**, *5*, 1–7. [CrossRef]
20. Sylvester, M.F.; Yannas, I.V.; Salzman, E.W.; Forbes, M.J. Collagen banded fibril structure and the collagen-platelet reaction. *Thromb Res.* **1989**, *55*, 135–148. [CrossRef] [PubMed]
21. Karabekiroğlu, S.; Kahraman, F.B.; Ünlü, N. Effectiveness of different preventive agents on initial occlusal and proximal caries lesions, A follow-up study. *J. Dent. Sci.* **2018**, *13*, 109–115. [CrossRef]
22. Kassa, D.; Day, P.; High, A.; Duggal, M. Histological comparison of pulpal inflammation in primary teeth with occlusal or proximal caries. *Int. J. Paediatr. Dent.* **2009**, *19*, 26–33. [CrossRef]
23. Pitts, N.B.; Ekstrand, K.; ICDAS Foundation. International Caries Detection and Assessment System (ICDAS) and its International Caries Classification and Management System (ICCMS)-methods for staging of the caries process and enabling dentists to manage caries. *Community Dent. Oral. Epidemiol.* **2013**, *41*, e41–e52. [CrossRef] [PubMed]
24. Fuentes, V.; Toledano, M.; Osorio, R.; Carvalho, R.M. Microhardness of superficial and deep sound human dentin. *J. Biomed. Mater. Res. Part A* **2003**, *66*, 850–853. [CrossRef] [PubMed]
25. Kinoshita, H.; Miyoshi, N.; Fukunaga, Y.; Ogawa, T.; Ogasawara, T.; Sano, K. Functional mapping of carious enamel in human teeth with Raman microspectroscopy. *J. Raman Spectrosc.* **2008**, *39*, 655–660. [CrossRef]
26. Almahdy, A.; Downey, F.C.; Sauro, S.; Cook, R.C.; Sherriff, M.; Richards, D.; Watson, T.F.; Banerjee, A.; Festy, F. Microbiochemical analysis of carious dentine using Raman and fluorescence spectroscopy. *Caries Res.* **2012**, *46*, 432–440. [CrossRef] [PubMed]
27. Liu, Y.; Tjäderhane, L.; Breschi, L.; Mazzoni, A.; Li, N.; Mao, J.; Pashley, D.H.; Tay, F.R. Limitations in bonding to dentin and experimental strategies to prevent bond degradation. *J. Dent. Res.* **2011**, *90*, 953–968. [CrossRef]
28. Linde, A. Dentin matrix proteins, composition and possible functions in calcification. *Anat Rec.* **1989**, *224*, 154–166. [CrossRef]
29. Ramakrishnaiah, R.; Rehman, G.U.; Basavarajappa, S.; Al Khuraif, A.A.; Durgesh, B.H.; Khan, A.S.; Rehman, I.U. Applications of Raman spectroscopy in dentistry, analysis of tooth structure. *Appl. Spectrosc. Rev.* **2015**, *50*, 332–350. [CrossRef]
30. Krafft, C.; Sergo, V. Biomedical applications of Raman and infrared spectroscopy to diagnose tissues. *Spectrosc* **2006**, *20*, 195–218. [CrossRef]
31. Maske, T.T.; Isolan, C.P.; van de Sande, F.H.; Peixoto, A.C.; Faria-e-Silva, A.L.; Cenci, M.S.; Moraes, R.R. A biofilm cariogenic challenge model for dentin demineralization and dentin bonding analysis. *Clin. Oral Investig.* **2015**, *19*, 1047–1053. [CrossRef]
32. Paschalis, E.P.; Verdelis, K.; Doty, S.B.; Boskey, A.L.; Mendelsohn, R.; Yamauch, M.I. Spectroscopic characterization of collagen cross-links in bone. *J. Bone Miner. Res.* **2001**, *16*, 1821–1828. [CrossRef]
33. Dirksen, T.R. Lipid components of sound and carious dentine. *J. Dent. Res.* **1963**, *42*, 128–132. [CrossRef]
34. Seredin, P.; Goloshchapov, D.; Prutskij, T.; Ippolitov, Y. Phase transformations in a human tooth tissue at the initial stage of caries. *PLoS ONE* **2015**, *10*, e0124008. [CrossRef]
35. Gourion-Arsiquaud, S.; West, P.A.; Boskey, A.L. Fourier transform-infrared microspectroscopy and microscopic imaging. *Osteoporosis* **2008**, *2*, 293–303. [CrossRef]

36. Pugach, M.; Strother, J.; Darling, C.; Fried, D.; Gansky, S.; Marshall, S.; Marshall, G. Dentin caries zones, mineral, structure, and properties. *J. Dent. Res.* **2009**, *88*, 71–76. [CrossRef] [PubMed]
37. Ami, D.; Natalello, A.; Gatti-Lafranconi, P.; Schultz, T.; Lotti, M.; de Marco, A.; Doglia, S.M. FT-IR spectroscopy for the study of bacterial membrane stress induced by recombinant protein production. *Microb. Cell Factories* **2006**, *5*, 1–2. [CrossRef]
38. Braga, M.M.; Floriano, I.; Ferreira, F.R.; Silveira, J.M.; Reyes, A.; Tedesco, T.K.; Raggio, D.P.; Imparato, J.C.; Mendes, F.M. Are the approximal caries lesions in primary teeth a challenge to deal with? A critical appraisal of recent evidences in this field. VIRDI, MC Emerging trends in Oral Health Sciences and Dentistry. *Intech* **2015**, *11*, 83–116. [CrossRef]
39. de Barros Pinto, L.; Lira, M.L.; Cavalcanti, Y.W.; Dantas, E.L.; Vieira, M.L.; de Carvalho, G.G.; de Sousa, F.B. Natural enamel caries, dentine reactions, dentinal fluid and biofilm. *Sci. Rep.* **2019**, *9*, 1–9. [CrossRef] [PubMed]
40. Shellis, R.P. *Transport Processes in Enamel and Dentine*; Tooth wear and sensitivity; Martin Dunitz: London, UK, 2000; pp. 19–28.
41. Banerjee, A.; Frencken, J.E.; Schwendicke, F.; Innes, N.P.T. Contemporary operative caries management, consensus recommendations on minimally invasive caries removal. *Br. Dent. J.* **2017**, *223*, 215–222. [CrossRef] [PubMed]

diagnostics

Article

# Diagnostic Accuracy of Line-Field Confocal Optical Coherence Tomography for the Diagnosis of Skin Carcinomas

**Elisa Cinotti [1,2,*], Tullio Brunetti [1], Alessandra Cartocci [1], Linda Tognetti [1], Mariano Suppa [2,3], Josep Malvehy [4,5], Javiera Perez-Anker [4,5], Susanna Puig [4,5], Jean Luc Perrot [6] and Pietro Rubegni [1]**

1 Department of Medical, Surgical and Neurological Sciences, Dermatology Section, University of Siena, 53100 Siena, Italy
2 Groupe d'Imagerie Cutanée Non Invasive (GICNI), Société Française de Dermatologie (SFD), 75008 Paris, France
3 Department of Dermatology, Hôpital Erasme, Université Libre de Bruxelles, 1050 Brussels, Belgium
4 Melanoma Unit, Hospital Clinic Barcelona, University of Barcelona, 08007 Barcelona, Spain
5 CIBER de Enfermedades Raras, Instituto de Salud Carlos III, 08007 Barcelona, Spain
6 Department of Dermatology, University Hospital of St-Etienne, 42270 Saint-Etienne, France
* Correspondence: elisa.cinotti@unisi.it; Tel.: +39-0577-585-428; Fax: +39-0577-585-484

**Abstract:** Line-field confocal optical coherence tomography (LC-OCT) is a new, noninvasive imaging technique for the diagnosis of skin cancers. A total of 243 benign (54%) and malignant (46%) skin lesions were consecutively enrolled from 27 August 2020, to 6 October 2021 at the Dermatology Department of the University Hospital of Siena, Italy. Dermoscopic- and LC-OCT-based diagnoses were given by an expert dermatologist and compared with the ground truth. Considering all types of malignant skin tumours (79 basal cell carcinomas (BCCs), 22 squamous cell carcinomas, and 10 melanomas), a statistically significant increase ($p = 0.013$) in specificity was observed from dermoscopy (0.73, CI 0.64–0.81) to LC-OCT (0.87, CI 0.79–0.93) while sensitivity was the same with the two imaging techniques (0.95 CI 0.89–0.98 for dermoscopy and 0.95 CI 0.90–0.99 for LC-OCT). The increase in specificity was mainly driven by the ability of LC-OCT to differentiate BCCs from other diagnoses. In conclusion, our real-life study showed that LC-OCT can play an important role in helping the noninvasive diagnosis of malignant skin neoplasms and especially of BCCs. LC-OCT could be positioned after the dermoscopic examination, to spare useless biopsy of benign lesions without decreasing sensitivity.

**Keywords:** optical coherence tomography; tumor; basal cell carcinoma; imaging; squamous cell carcinoma

**Citation:** Cinotti, E.; Brunetti, T.; Cartocci, A.; Tognetti, L.; Suppa, M.; Malvehy, J.; Perez-Anker, J.; Puig, S.; Perrot, J.L.; Rubegni, P. Diagnostic Accuracy of Line-Field Confocal Optical Coherence Tomography for the Diagnosis of Skin Carcinomas. *Diagnostics* **2023**, *13*, 361. https://doi.org/10.3390/diagnostics13030361

Academic Editor: Viktor Dremin

Received: 23 December 2022
Revised: 9 January 2023
Accepted: 12 January 2023
Published: 18 January 2023

**Copyright:** © 2023 by the authors. Licensee MDPI, Basel, Switzerland. This article is an open access article distributed under the terms and conditions of the Creative Commons Attribution (CC BY) license (https://creativecommons.org/licenses/by/4.0/).

## 1. Introduction

Line-field confocal optical coherence tomography (LC-OCT) is a new, noninvasive skin imaging technique that combines the advantages of optical coherence tomography (OCT) and reflectance confocal microscopy (RCM) in terms of spatial resolution, penetration, and image orientation, overcoming their respective limits [1–5]. LC-OCT has a higher resolution than OCT (~1 μm) [6–8] and higher penetration [9] depth than RCM (~500 μm), and it creates both vertical and horizontal images in real time [10–13].

Recently, LC-OCT [14] has been gaining attention because it has been shown that this device can help the clinical diagnosis of different neoplastic [15–18], inflammatory [19–23], and infectious [24,25] skin diseases. In particular, it has proven to be very effective in identifying basal cell carcinoma (BCC) [26], even managing to differentiate its histological subtypes [27] and to follow-up noninvasive treatment [28]. Furthermore, this noninvasive diagnostic technique can be used to help the differentiation of actinic keratosis (AK) from squamous cell carcinoma (SCC) [29–31] and to monitor the field of cancerization-directed treatments [32].

Many descriptive studies have shown the relevance of this imaging device for the diagnosis of cutaneous tumours, and our study aimed to evaluate the sensitivity and specificity of LC-OCT compared to dermoscopy for the diagnosis of skin tumours in a real-life setting in a third-level dermatology department.

## 2. Materials and Methods

### 2.1. Study Design

Prospective observational, monocentric study.

### 2.2. Setting

Patients were enrolled from the 27 August 2020 to the 6 October 2021 at the Dermatology Department of the University Hospital of Siena, Italy, from the melanoma prevention outpatient ambulatory. The study was conducted according to the criteria set by the Declaration of Helsinki. All data were deidentified before use.

### 2.3. Participants

We enrolled consecutive patients with cutaneous lesions of clinical and/or dermoscopic uncertain diagnosis of possible malignant skin tumours that needed to be removed or followed up according to a skin imaging expert dermatologist (E.C.) and that had LC-OCT examination.

### 2.4. Imaging Examination

Dermoscopy was performed both with a hand-held 10× dermoscope (DermLite DL4, DermLite, San Juan Capistrano, USA) and a 20× videodermoscope (Vivacam, Mavig, Munich, Germany). LC-OCT (DeepLive, Damae, France) was performed in horizontal and vertical scans and 3D mode, and a video was acquired on the operator's judgment. Following manufacturer recommendations, lesions in the patients' periocular region were excluded from LC-OCT examination.

Dermoscopic and LC-OCT diagnoses were given by an expert in skin imaging (E.C.) during the imaging examination of the lesions and were registered on the LC-OCT device. Concerning LC-OCT, BCC was diagnosed by the presence of tumour lobules [27], SCC by the presence of atypical keratinocytes in the entire epidermis [31,32], and melanoma by the presence of atypical bright cells that were sparse inside the epidermis and/or inside melanocytic nests [12–15]. The lesions suggesting malignant skin tumours at dermoscopic and/or LC-OCT examination were biopsied or surgically removed for histological diagnosis. The others were followed up for at least one year.

### 2.5. Statistical Analysis

The sensitivity and specificity of each technique (LC-OCT and dermoscopy) for the diagnosis of BCC, SCC/Bowen disease (i.e., SCC or Bowen disease), AK/SCC/Bowen disease (i.e., AK or SCC or Bowen disease) group and malignant tumour were calculated with their exact 95% confidence interval (CI) by using the histopathological diagnosis obtained from an incisional or excisional biopsy as the gold standard; a distinct analysis was also performed, including the lesions that had a final diagnosis based on a follow up of at least one year without a histological examination.

The sensitivity and specificity of LC-OCT and dermoscopy for the diagnosis of the different skin tumours were compared by using the proportion test. $p$ values $< 0.05$ were considered statistically significant. All statistical analyses were conducted by using R (version 4.0.3., R foundation for statistical computing).

## 3. Results

We included 196 patients (81 women, 115 men; mean age of 64.45 years, range 0–96 years) with 243 lesions; 226 lesions were histopathologically confirmed (Table 1) and 17 lesions had a final diagnosis after a follow-up of at least one year. Sensitivity and specificity of dermoscopy and LC-OCT for BCC, SCC/Bowen disease group, AK/SCC/Bowen disease group, and malignant tumour considering histopathology as the gold standard are reported in Table 2.

**Table 1.** Confusion matrix: dermoscopy vs. histology and LC OCT vs. histology.

| | | HISTOLOGY | | | | | | | |
|---|---|---|---|---|---|---|---|---|---|
| | | BCC (n = 79) | Benign ML (n = 22) | Melanoma (n = 10) | AK (n = 16) | SCC (n = 22) | Inflammatory Lesion(n = 14) | Rare Disease (n = 5) | Other (n = 58) |
| DERMOSCOPY (in case of multiple diagnoses on dermoscopy, the worst diagnosis was retained) | BCC (n = 96) | 76 | 2 | 0 | 1 | 2 | 3 | 0 | 12 |
| | Benign ML (n = 17) | 0 | 15 | 1 | 0 | 0 | 0 | 0 | 1 |
| | Melanoma (n = 14) | 0 | 5 | 9 | 0 | 0 | 0 | 0 | 0 |
| | AK (n = 15) | 1 | 0 | 0 | 12 | 1 | 0 | 0 | 1 |
| | SCC (n = 26) | 1 | 0 | 0 | 3 | 17 | 2 | 0 | 3 |
| | Inflammatory lesion (n = 6) | 0 | 0 | 0 | 0 | 0 | 5 | 0 | 1 |
| | Rare disease (n = 5) | 0 | 0 | 0 | 0 | 0 | 0 | 5 | 0 |
| | Other (n = 47) | 1 | 0 | 0 | 0 | 2 | 4 | 0 | 40 |
| LC-OCT (in case of multiple diagnoses on LC-OCT, the worst diagnosis was retained) | BCC (n = 84) | 77 | 1 | 0 | 0 | 0 | 1 | 0 | 5 |
| | benign ML (n = 20) | 0 | 17 | 1 | 0 | 0 | 0 | 0 | 2 |
| | Melanoma (n = 13) | 0 | 3 | 9 | 0 | 0 | 0 | 0 | 1 |
| | AK (n = 18) | 1 | 0 | 0 | 14 | 2 | 0 | 0 | 1 |
| | SCC (n = 24) | 1 | 0 | 0 | 1 | 19 | 1 | 0 | 2 |
| | Inflam (n = 9) | 0 | 0 | 0 | 0 | 0 | 8 | 0 | 1 |
| | rare disease (n = 5) | 0 | 0 | 0 | 0 | 0 | 0 | 5 | 0 |
| | Other (n = 53) | 0 | 1 | 0 | 1 | 1 | 4 | 0 | 46 |

**Table 2.** Sensitivity and specificity of dermoscopy and LC-OCT considering only histologically confirmed cases.

| | | DERMOSCOPY | LC-OCT | *p*-Value |
|---|---|---|---|---|
| BCC (n = 79) | TP/P | 76/79 | 77/79 | |
| | TN/N | 127/147 | 140/147 | |
| | Sensitivity (CI) | 0.96 (0.89–0.99) | 0.97 (0.91–1.00) | 1 |
| | Specificity (CI) | 0.86 (0.80–0.91) | 0.95 (0.90–0.98) | 0.015 |
| SCC/Bowen (n = 19) | TP/P | 17/22 | 19/22 | |
| | TN/N | 195/204 | 199/204 | |
| | Sensitivity (CI) | 0.77 (0.55–0.92) | 0.86 (0.65–0.97) | 0.696 |
| | Specificity (CI) | 0.96 (0.92–0.98) | 0.98 (0.94–0.99) | 0.415 |
| AK/Bowen/SCC (n = 36) | TP/P | 33/38 | 36/38 | |
| | TN/N | 180/188 | 182/188 | |
| | Sensitivity (CI) | 0.87 (0.72–0.96) | 0.95 (0.82–0.99) | 0.428 |
| | Specificity (CI) | 0.96 (0.92–0.98) | 0.97 (0.93–0.99) | 0.785 |
| Malignant vs. non Malignant (n = 111) | TP/P | 105/111 | 106/111 | |
| | TN/N | 84/115 | 100/115 | |
| | Sensitivity (CI) | 0.95 (0.89–0.98) | 0.95 (0.90–0.99) | 1 |
| | Specificity (CI) | 0.73 (0.64–0.81) | 0.87 (0.79–0.93) | 0.013 |

TP, true positive; P, positive; TN, true negative; N, negative.

### 3.1. Diagnostic Performances of Dermoscopy and LC-OCT Considering Only Cases with Histological Diagnoses

#### 3.1.1. Dermoscopy and LC-OCT Diagnostic Performances for BCC

Considering the 79 histopathologically confirmed BCCs, LC-OCT showed higher specificity (0.95, CI 0.90–0.98; $p = 0.015$) for BCC diagnosis than dermoscopy (0.86, CI 0.80–0.91, $p = 0.015$), and no statistically significant difference in sensitivity (0.97 CI 0.91–1.00 for LC-OCT and 0.96 CI 0.89–0.99 for dermoscopy; Table 2). Dermoscopy had 20 false positive (FP) cases that histologically corresponded to two nevi, one AK, two Bowen diseases in situ, three inflammatory lesions, a scar, and 11 benign nonmelanocytic lesions (including a solar lentigo, a seborrheic keratosis (SK), a lichenoid keratosis, a xanthogranuloma, a sebaceoma, a trichilemmoma, a trichoblastoma, and a neurofibroma). LC-OCT had seven FP cases that histologically corresponded to one nevus, one inflammatory lesion, and five benign nonmelanocytic lesions. LC-OCT enabled us to correctly diagnose 13 of the 20 dermoscopic FP BCCs as four inflammatory lesions, two cases of normal skin, two Bowen diseases, one AK, one nevus, one neurofibroma, one scar, and one xanthogranuloma. Three dermoscopic false negative (FN) cases were diagnosed at dermoscopy as AK, SCC, and SK and histologically corresponded to two infiltrative BCCs and a superficial microinvasive BCC. Two of these FN cases were also FN at LC-OCT and were diagnosed as AK and SCC, while they corresponded to two infiltrative BCCs at histopathology.

#### 3.1.2. Dermoscopy and LC-OCT Diagnostic Performances for the Diagnosis of SCC/Bowen Disease

LC-OCT showed a slightly higher sensitivity (0.86, CI 0.65–0.97) for SCC/Bowen disease diagnosis than dermoscopy (0.77, CI 0.55–0.92), which did not reach a statistically significant difference. Concerning specificity, no statistically significant difference was found (0.98, CI 0.94–0.99 for LC-OCT and 0.96,0.92–0.98 for dermoscopy; Table 2). Among the 22 histopathologically confirmed SCC/Bowen diseases, there were nine FP at dermoscopy, five FP at LC-OCT, five FN on dermoscopy, and three FN on LC-OCT (two of them were Bowen diseases in situ diagnosed as AK). The nine FP at dermoscopy were histologically diagnosed as one BCC, three AKs, two inflammatory lesions, one dermatofibroma, one granulomatous lesion, and one microcystic adnexal carcinoma; the five FP at LC-OCT were histologically diagnosed as one BCC, one AK, one inflammatory lesion, one dermatofibroma, and one granulomatous lesion.

The five FN at dermoscopy were diagnosed as one AK, two BCCs, one SK, and one granuloma, and corresponded to four Bowen diseases in situ and one microinvasive keratoacanthoma at histopathology; the three FN at LC-OCT were diagnosed as two AKs and one SK and corresponded to two Bowen diseases in situ and a microinvasive keratoacanthoma at histopathology.

#### 3.1.3. Dermoscopy and LC-OCT Diagnostic Performances for the Diagnosis of AK/SCC/Bowen Disease

LC-OCT showed higher sensitivity and specificity than dermoscopy for AK/SCC/Bowen disease diagnosis (sensitivity of 0.87 (CI 0.72–0.96) for dermoscopy and 0.95 (CI 0.82–0.99) for LC-OCT, specificity of 0.96 (CI 0.92–0.98) for dermoscopy and 0.97 (CI 0.93–0.99) for LC-OCT (Table 2)). However, the difference in sensitivity and specificity did not reach statistical significance.

#### 3.1.4. Dermoscopy and LC-OCT Diagnostic Performances for Malignant Tumour

Considering only the cases with a histologic diagnosis of malignancy, we observed a significant increase of specificity from 0.73 (CI 0.64–0.81) with dermoscopy to 0.87 (CI 0.79–0.93) with LC-OCT ($p = 0.013$) for a malignant tumour, whereas the sensitivity was similar with the two imaging techniques (0.95 CI 0.89–0.98 for dermoscopy and 0.95 CI 0.90–0.99 for LC-OCT). The group of malignant tumours included both skin carcinomas and melanomas.

#### 3.1.5. Diagnostic Performances of Dermoscopy and LC-OCT Considering Both Histological and Follow-Up Diagnoses

The sensitivity and specificity of dermoscopy and LC-OCT for BCC and malignant tumour considering as comparison the diagnoses derived from histopathology and follow-up at least one year are reported in Table 3.

**Table 3.** Sensitivity and specificity of dermoscopy and LC-OCT considering histology and follow-up diagnoses.

| | | DERMOSCOPY | LC-OCT | *p* Values |
|---|---|---|---|---|
| **BCC (n = 79)** | **TP/P** | 76/79 | 77/79 | |
| | **TN/N** | 127/160 | 153/160 | |
| | **Sensitivity (CI)** | 0.96 (0.89–0.99) | 0.97 (0.91–1.00) | 1 |
| | **Specificity (CI)** | 0.79 (0.72–0.85) | 0.96 (0.91–0.98) | $p < 0.001$ |
| **Malignant vs. non Malignant (n = 111)** | **TP/P** | 105/111 | 106/111 | |
| | **TN/N** | 84/132 | 117/132 | |
| | **Sensitivity (CI)** | 0.95 (0.89–0.98) | 0.95 (0.90–0.99) | 1 |
| | **Specificity (CI)** | 0.64 (0.55–0.72) | 0.89 (0.82–0.93) | $p < 0.001$ |

#### 3.1.6. Dermoscopy and LC-OCT Diagnostic Performances for BCC (Including 13 Cases without a Histological Diagnosis)

Considering both the cases with histology and follow-up of at least one year, LC-OCT enabled us to correctly diagnose 26 over 33 dermoscopic FP cases: seven inflammatory lesions, three cases of normal skin, three scars, two Bowenoid SCCs in situ, two AKs, three nevi, one neurofibroma, one xanthogranuloma, one seborrhoeic keratosis, one rosacea, one sebaceous hyperplasia, and one scaly crust with papillomatosis. Among these cases, LC-OCT allowed us to save 13 excisions. Considering these 13 FP lesions at dermoscopy for which BCC was excluded after LC-OCT and for which surgical excision was not done (assuming that the follow-up >1 year of these patients could confirm the absence of BCC), the specificity for BCC diagnosis increased from 0.79 (CI 0.72–0.85) for dermoscopy to 0.96 (CI 0.91–0.98) for LC-OCT ($p < 0.001$). Sensitivity was similar for LC-OCT (0.97, CI 0.91) and dermoscopy (0.96, CI 0.89–0.99, Table 3).

#### 3.1.7. Dermoscopy and LC-OCT Diagnostic Performances for Malignant Tumours (Including 17 Cases without a Histological Diagnosis)

Considering the 17 FP lesions at dermoscopy for which malignancy was excluded after LC-OCT and surgical excision was not done (assuming that the follow-up of these patients could confirm the absence of malignant tumour), the specificity for malignancy increased respectively from 0.64 (CI 0.55–0.72) of dermoscopy to 0.89 (CI 0.82–0.93, $p < 0.001$) for LC-OCT. Sensitivity was similar: 0.95 (CI 0.90–0.99) for LC-OCT and 0.95 (CI 0.89–0.98) for dermoscopy (Table 3).

## 4. Discussion

Our study showed that LC-OCT can increase specificity for the noninvasive diagnosis of skin cancers compared to dermoscopy. Considering histopathology as a gold standard and analyzing only the cases with histological diagnosis (Table 2), we found an increase in specificity for the diagnosis of BCC from 0.86 (0.80–0.91 CI) for dermoscopy to 0.95 (0.90–0.98 CI) for LC-OCT ($p = 0.015$). The sensitivity was similar with the two methods (0.96 CI 0.89–0.99 for dermoscopy with three FN cases and 0.97 CI 0.91–1.00 for LC-OCT with two FN cases).

The same analysis including the cases that were diagnosed based on a follow-up of at least one year and that lacked histopathological examination obtained similar results. We found an increase in the specificity for the diagnosis of BCC from 0.79 (CI 0.72–0.85) for dermoscopy to 0.96 (CI 0.91–0.98) for LC-OCT ($p < 0.001$), whereas with regard to sensitivity we did not find any statistically significant difference between dermoscopy (0.96 CI 0.89–0.99) and LC-OCT (0.97 CI 0.91–1.00). Similar sensitivity results probably reflect the current use of LC-OCT as a secondary-level technique on skin lesions that are already identified as suspicious by dermoscopic examination. In most cases, LC-OCT easily confirms the dermoscopic diagnosis of a malignant tumour [27] and it is interesting to note that in clinical practice, LC-OCT is useful to increase the diagnostic confidence of the dermatologist and to confirm the need for surgical excision.

LC-OCT only missed two infiltrative BCCs, and retrospective examination of their images revealed hyperkeratosis thickness ranging from 200 to 300 μm with no visible dermis in one case and poor image in the other case

Concerning specificity, LC-OCT significantly reduced the cases of false positives (FP) BCCs in our series. 13 FP cases of BCC at dermoscopy were correctly diagnosed with LC-OCT (Table 3; Figures 1 and 2). These data are consistent with the latest studies on LC-OCT that highlight how this technique can easily recognize BCC imitators [20,33–35].

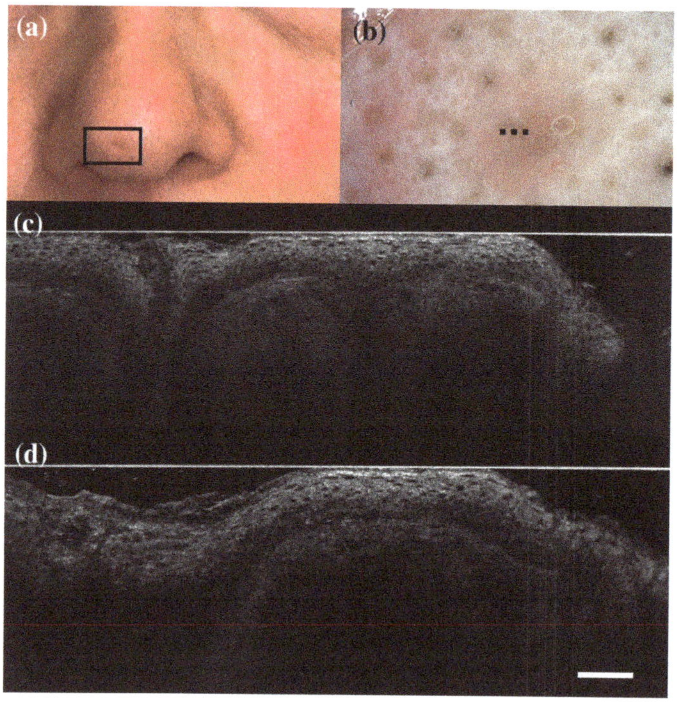


**Figure 1.** False positive case of basal cell carcinoma at dermoscopy. Clinical (**a**), dermoscopic (**b**), and LC-OCT images (**c**,**d**). Dermoscopy identified the lesion as a basal cell carcinoma, while LC-OCT as healthy skin. Histology confirmed the diagnosis of healthy skin. Dashed line (**b**) indicates the approximate area of the LC-OCT imaging. White scale bar in LC-OCT images: 100 μm.

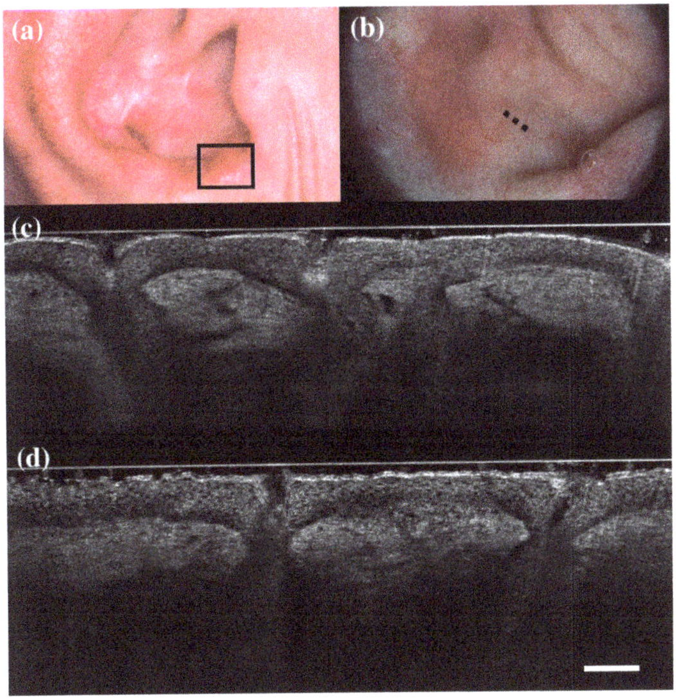


**Figure 2.** False positive case of basal cell carcinoma at dermoscopy. Clinical (**a**), dermoscopic (**b**), and LC-OCT images (**c**,**d**). Dermoscopy identified the lesion as a basal cell carcinoma, while LC-OCT identified it as healthy skin. Histology confirmed the diagnosis of healthy skin. Dashed line (**b**) indicates the approximate area of the LC-OCT imaging. White scale bar in LC-OCT images: 100 μm.

There were only seven FP cases at LC-OCT, and three of them corresponded to benign skin tumours that can share histopathological similarities with BCC: a sebaceoma, a trichoblastoma, and a trichilemmoma. Sebaceoma (Figure 3) is characterized by multiple basal cell nests (Figure 3, asterisk) with a random mix of sebaceous cells in the upper and middle dermis with possible continuity with the basal layer of the epidermis [36]. Trichoblastoma (Figure 4) shows irregular nests of basal cells similar to BCC, with variable stromal thickening and pilar differentiation [36]. Trichilemmoma (Figure 5) is composed of one or more lobules (Figure 5, asterisk) in the dermis that extend in continuity with the epidermis (Figure 5, orange arrow) or the follicular epithelium, and there is a peripheral layer of columnar palisade cells [36]. Under LC-OCT, all these three tumours exhibited tumour islands with overlapping features of BCC. We should also consider that there could be biopsy sampling errors explaining some FP results. Interestingly, one LC-OCT FP case of our series, defined on the histological report of an incisional biopsy as solar lentigo, was later completely excised based on the retrospective revaluation of the LC-OCT images and had a final histological diagnosis of BCC.

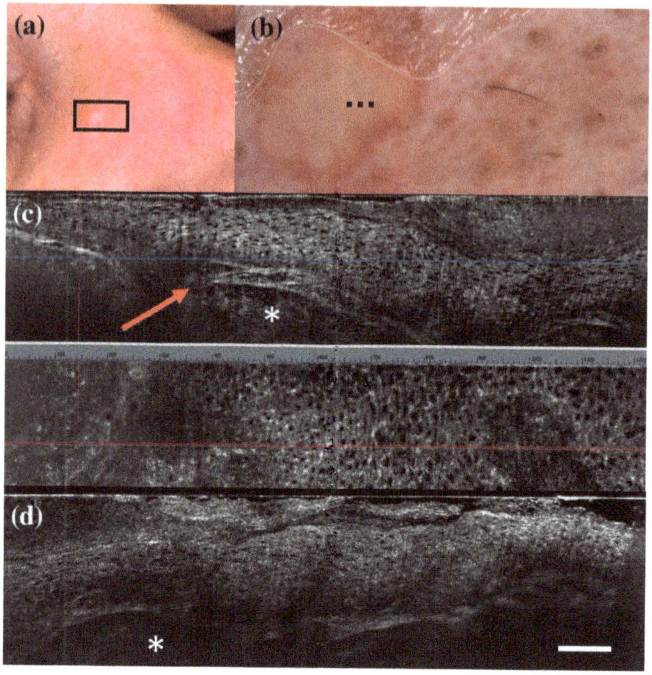


**Figure 3.** Sebaceoma diagnosed as basal cell carcinoma at dermoscopy and line-field confocal optical coherence tomography (LC-OCT). Clinical (**a**), dermoscopic (**b**), and LC-OCT images (**c**,**d**). Dermoscopy shows a pinkish-yellowish background and linear vessels. LC-OCT reveals a large lobular structure (asterisk) with "feuilletage" and clefting, surrounded by hyperreflective stroma and connected to a hair follicle (red arrow). Dashed line (**b**) indicates the approximate area of the LC-OCT imaging. White scale bar in LC-OCT images: 100 µm.

In the literature, we could find only one prospective study on the diagnostic accuracy of LC-OCT for skin tumours, and it consists of similar real-life research on equivocal lesions. It showed that LC-OCT can significantly increase diagnostic confidence after dermoscopy and avoid potentially unnecessary biopsies [37]. However, it revealed a higher sensitivity (0.98) and a good, but lower, specificity (0.80) for LC-OCT compared to dermoscopy (sensitivity of 0.90 and specificity of 0.86).

The acquisition and interpretation of the LC-OCT images are operator-dependent and different results may be related to different investigator expertise and different lesion selection (equivocal aspect of the lesion at dermoscopy). Moreover, Gust et al. found that in 70% of the lesions, the LC-OCT diagnostic was provided with high confidence in comparison with dermoscopy which only provided high confidence in 48% of the lesions [37]. In this subgroup, the LC-OCT performance increased significantly, with a sensitivity of 100% and a specificity of 97% in agreement with our results. In the future, an effort should be made to define precise criteria for the LC-OCT diagnosis of skin tumours to have more reliable and comparable results. Moreover, artificial intelligence could help the identification of BCC tumour lobules and atypical cells [10]. Regarding the diagnosis of SCC/Bowen's disease/AK, we found an increase in both sensitivity and specificity for LC-OCT compared to dermoscopy without any statistically significant difference. Recently, many studies have shown that LC-OCT can identify several histological criteria of AK and SCC and this technique seems to be promising for the diagnosis of squamous cell tumours [16,29,31,32]. Our study could not prove a statistically significant benefit of LC-OCT possibly due to the relatively small sample size that has been analyzed.

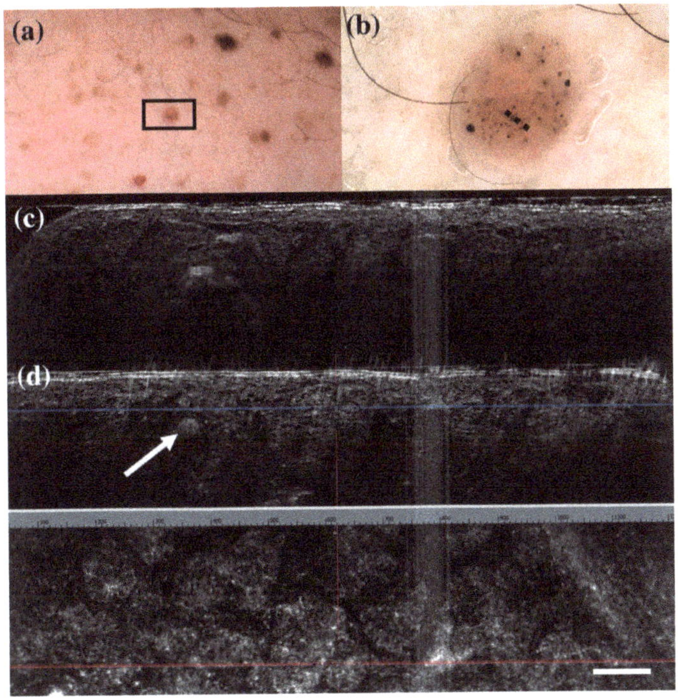

**Figure 4.** Trichoblastoma diagnosed as basal cell carcinoma under dermoscopy and line-field confocal optical coherence tomography (LC-OCT). Clinical (**a**), dermoscopic (**b**), and LC-OCT images (**c**,**d**). Dermoscopy shows black, brown, and grey globules and dots on a pinkish and brownish background. LC-OCT reveals small well-delimited roundish lobules with "palisading" and some keratin cysts (white arrow). Dashed line (**b**) indicates the approximate area of the LC-OCT imaging. White scale bar in LC-OCT images: 100 µm.

Regarding the diagnosis of malignancy, considering histopathology as a gold standard and analyzing only cases with histological diagnosis, a significant increase in specificity (Table 2) was observed from 0.73 (IC 0.64–0.81) with dermoscopy to 0.87 (IC 0.79–0.93) with LC-OCT ($p = 0.013$). However, we did not detect any statistically significant difference in sensitivity between the two methods (0.95 IC 0.89–0.98 for dermoscopy and 0.95 IC 0.90–0.99 for LC-OCT), similar to the diagnosis of BCCs. The same analysis including the 17 FP cases of malignity at dermoscopy in which the diagnosis of malignant neoplasm was excluded with LC-OCT and surgery was not performed (assuming that the follow-up of these patients could confirm the absence of malignant tumour, (see Table 3)) obtained similar results. We detected an increase in specificity for malignant skin tumours from 0.64 (CI 0.55–0.72) for dermoscopy to 0.89 (CI 0.82–0.93) for LC-OCT ($p << 0.001$) while regarding the sensitivity we did not find any statistically significant difference between dermoscopy (0.96 CI 0.89–0.99) and LC-OCT (0.97 CI 0.91–1.00). These data on malignant tumours were mainly driven by BCCs and SCCs because melanomas were few. The increase in specificity with LC-OCT for the diagnosis of malignant skin tumours was mainly determined by the increase in specificity for the diagnosis of BCC. Although LC-OCT seems to play a possible role also for melanocytic tumours [12], to date there are few data on the diagnostic accuracy of LC-OCT for malignant skin tumours other than BCC and SCC [15].

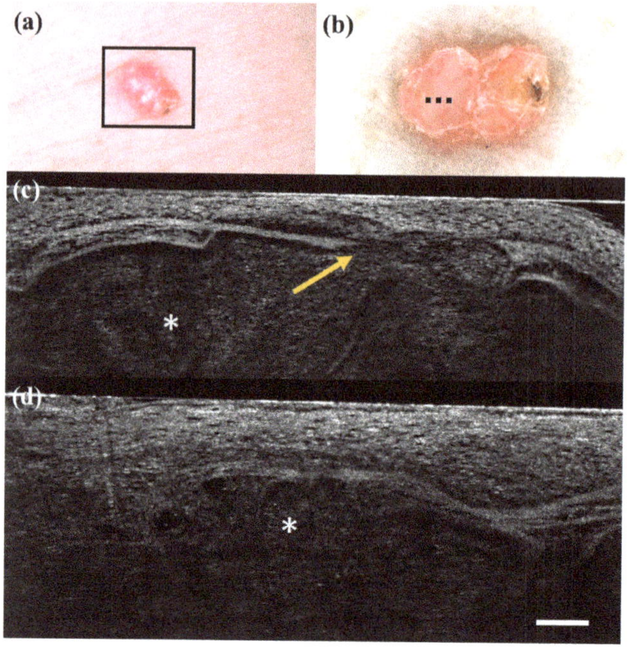


**Figure 5.** Trichillemoma diagnosed as basal cell carcinoma under dermoscopy and line-field confocal optical coherence tomography (LC-OCT). Clinical (**a**), dermoscopic (**b**), and LC-OCT images (**c**,**d**). Dermoscopy shows a pinkish background with linear vessels and scales. LC-OCT reveals lobular structures (asterisk) with clefting, palisade, "feuilletage", and connection with the epidermis (orange arrow). Dashed line (**b**) indicates the approximate area of the LC-OCT imaging. White scale bar in LC-OCT images: 100 μm.

## 5. Conclusions

- Our real-life study confirmed that dermoscopy can select lesions at risk of being malignant skin tumours (very sensitive tool).
- LC-OCT could be positioned in a second line to rule out malignancy to spare useless biopsy without decreasing sensitivity (very specific tool).
- LC-OCT can help in the identification of BCC with only 10 diagnostic errors in our entire database covering more than one year.
- LC-OCT seems to also be promising for keratinocyte tumours (AK, SCC, and Bowen's disease) by increasing the specificity and reducing FP cases compared to dermoscopy.
- Further studies should be performed to confirm our data and investigate the possible role of LC-OCT for the different malignant skin tumours.

**Author Contributions:** Conceptualization, E.C. and J.L.P.; Methodology, E.C.; Software, A.C.; Validation, E.C.; Formal analysis, E.C. and A.C.; Investigation, E.C.; Resources, E.C.; Data curation, E.C., T.B. and A.C.; Writing—original draft, E.C. and T.B.; Writing—review & editing, E.C., L.T., M.S., J.M., J.P.-A., S.P., J.L.P. and P.R. All authors have read and agreed to the published version of the manuscript.

**Funding:** This research received no external funding.

**Institutional Review Board Statement:** Ethical review and approval were waived for this study because this is on observational study and skin lesions were removed according to our clinical practice independently from this study.

**Informed Consent Statement:** Patient consent was waived due to the observational study type and data anonymization. Patient consent for image publication was obtained.

**Data Availability Statement:** Data are unavailable due to privacy restriction.

**Conflicts of Interest:** The authors declare no conflict of interest.

## References

1. Dubois, A.; Levecq, O.; Azimani, H.; Siret, D.; Barut, A.; Suppa, M.; Del Marmol, V.; Malvehy, J.; Cinotti, E.; Rubegni, P.; et al. Line-Field Confocal Optical Coherence Tomography for High-Resolution Noninvasive Imaging of Skin Tumors. *J. Biomed. Opt.* **2018**, *23*, 1–9. [CrossRef] [PubMed]
2. Davis, A.; Levecq, O.; Azimani, H.; Siret, D.; Dubois, A. Simultaneous Dual-Band Line-Field Confocal Optical Coherence Tomography: Application to Skin Imaging. *Biomed. Opt. Express* **2019**, *10*, 694–706. [CrossRef] [PubMed]
3. Tognetti, L.; Cinotti, E.; Falcinelli, F.; Miracco, C.; Suppa, M.; Perrot, J.-L.; Rubegni, P. Line-Field Confocal Optical Coherence Tomography: A New Tool for Non-Invasive Differential Diagnosis of Pustular Skin Disorders. *J. Eur. Acad. Dermatol. Venereol.* **2022**, *36*, 1873–1883. [CrossRef]
4. Gallay, C.; Ventéjou, S.; Gaide, O.; Christen-Zaech, S. Cutaneous visualization by different non-invasive skin imaging methods. *Rev. Med. Suisse* **2021**, *17*, 624–629.
5. Chauvel-Picard, J.; Bérot, V.; Tognetti, L.; Orte Cano, C.; Fontaine, M.; Lenoir, C.; Pérez-Anker, J.; Puig, S.; Dubois, A.; Forestier, S.; et al. Line-Field Confocal Optical Coherence Tomography as a Tool for Three-Dimensional in Vivo Quantification of Healthy Epidermis: A Pilot Study. *J. Biophotonics* **2021**, *15*, e202100236. [CrossRef] [PubMed]
6. Ruini, C.; Schuh, S.; Sattler, E.; Welzel, J. Line-Field Confocal Optical Coherence Tomography-Practical Applications in Dermatology and Comparison with Established Imaging Methods. *Skin Res. Technol.* **2021**, *27*, 340–352. [CrossRef]
7. Monnier, J.; Tognetti, L.; Miyamoto, M.; Suppa, M.; Cinotti, E.; Fontaine, M.; Perez, J.; Orte Cano, C.; Yélamos, O.; Puig, S.; et al. In Vivo Characterization of Healthy Human Skin with a Novel, Non-Invasive Imaging Technique: Line-Field Confocal Optical Coherence Tomography. *J. Eur. Acad. Dermatol. Venereol.* **2020**, *34*, 2914–2921. [CrossRef]
8. Ogien, J.; Levecq, O.; Azimani, H.; Dubois, A. Dual-Mode Line-Field Confocal Optical Coherence Tomography for Ultrahigh-Resolution Vertical and Horizontal Section Imaging of Human Skin in Vivo. *Biomed. Opt. Express* **2020**, *11*, 1327–1335. [CrossRef] [PubMed]
9. Schuh, S.; Ruini, C.; Sattler, E.; Welzel, J. Confocal line-field OCT. *Hautarzt Z. Dermatol. Venerol. Verwandte Geb.* **2021**, *72*, 1039–1047. [CrossRef]
10. Fischman, S.; Pérez-Anker, J.; Tognetti, L.; Di Naro, A.; Suppa, M.; Cinotti, E.; Viel, T.; Monnier, J.; Rubegni, P.; Del Marmol, V.; et al. Non-Invasive Scoring of Cellular Atypia in Keratinocyte Cancers in 3D LC-OCT Images Using Deep Learning. *Sci. Rep.* **2022**, *12*, 481. [CrossRef]
11. Xue, W.; Ogien, J.; Bulkin, P.; Coutrot, A.-L.; Dubois, A. Mirau-Based Line-Field Confocal Optical Coherence Tomography for Three-Dimensional High-Resolution Skin Imaging. *J. Biomed. Opt.* **2022**, *27*, 086002. [CrossRef]
12. Perez-Anker, J.; Puig, S.; Alos, L.; García, A.; Alejo, B.; Cinotti, E.; Orte Cano, C.; Tognetti, L.; Lenoir, C.; Monnier, J.; et al. Morphologic Evaluation of Melanocytic Lesions with Three-Dimensional Line-Field Confocal Optical Coherence Tomography: Correlation with Histopathology and Reflectance Confocal Microscopy. A Pilot Study. *Clin. Exp. Dermatol.* **2022**, *47*, 2222–2233. [CrossRef] [PubMed]
13. Waszczuk, L.; Ogien, J.; Perrot, J.-L.; Dubois, A. Co-Localized Line-Field Confocal Optical Coherence Tomography and Confocal Raman Microspectroscopy for Three-Dimensional High-Resolution Morphological and Molecular Characterization of Skin Tissues Ex Vivo. *Biomed. Opt. Express* **2022**, *13*, 2467–2487. [CrossRef]
14. Dubois, A.; Xue, W.; Levecq, O.; Bulkin, P.; Coutrot, A.-L.; Ogien, J. Mirau-Based Line-Field Confocal Optical Coherence Tomography. *Opt. Express* **2020**, *28*, 7918–7927. [CrossRef] [PubMed]
15. Schuh, S.; Ruini, C.; Perwein, M.K.E.; Daxenberger, F.; Gust, C.; Sattler, E.C.; Welzel, J. Line-Field Confocal Optical Coherence Tomography: A New Tool for the Differentiation between Nevi and Melanomas? *Cancers* **2022**, *14*, 1140. [CrossRef] [PubMed]
16. Cappilli, S.; Cinotti, E.; Lenoir, C.; Tognetti, L.; Perez-Anker, J.; Rubegni, P.; Puig, S.; Malvehy, J.; Perrot, J.L.; Del Marmol, V.; et al. Line-Field Confocal Optical Coherence Tomography of Basosquamous Carcinoma: A Case Series with Histopathological Correlation. *J. Eur. Acad. Dermatol. Venereol.* **2022**, *36*, 1214–1218. [CrossRef]
17. Miyachi, K.; Murakami, Y.; Inoue, Y.; Yoshioka, H.; Hirose, O.; Yamada, T.; Hasegawa, S.; Arima, M.; Iwata, Y.; Sugiura, K.; et al. UVA Causes Dysfunction of ETBR and BMPR2 in Vascular Endothelial Cells, Resulting in Structural Abnormalities of the Skin Capillaries. *J. Dermatol. Sci.* **2022**, *105*, 121–129. [CrossRef] [PubMed]
18. Tognetti, L.; Carraro, A.; Lamberti, A.; Cinotti, E.; Suppa, M.; Luc Perrot, J.; Rubegni, P. Kaposi Sarcoma of the Glans: New Findings by Line Field Confocal Optical Coherence Tomography Examination. *Skin Res. Technol.* **2021**, *27*, 285–287. [CrossRef]
19. Tognetti, L.; Cinotti, E.; Suppa, M.; Guazzo, R.; Habougit, C.; Santi, F.; Diet, G.; Fontaine, M.; Berot, V.; Monnier, J.; et al. Line Field Confocal Optical Coherence Tomography: An Adjunctive Tool in the Diagnosis of Autoimmune Bullous Diseases. *J. Biophotonics* **2021**, *14*, e202000449. [CrossRef]
20. Lacarrubba, F.; Verzì, A.E.; Puglisi, D.F.; Broggi, G.; Caltabiano, R.; Micali, G. Line-Field Confocal Optical Coherence Tomography of Xanthogranuloma: Correlation with Vertical and Horizontal Histopathology. *J. Cutan. Pathol.* **2021**, *48*, 1208–1211. [CrossRef]

21. Tognetti, L.; Carraro, A.; Cinotti, E.; Suppa, M.; Del Marmol, V.; Perrot, J.L.; Rubegni, P. Line-Field Confocal Optical Coherence Tomography for Non-Invasive Diagnosis of Lichenoid Dermatoses of the Childhood: A Case Series. *Skin Res. Technol.* **2021**, *27*, 1178–1181. [CrossRef]
22. Tognetti, L.; Rizzo, A.; Fiorani, D.; Cinotti, E.; Perrot, J.L.; Rubegni, P. New Findings in Non-Invasive Imaging of Aquagenic Keratoderma: Line-Field Optical Coherence Tomography, Dermoscopy and Reflectance Confocal Microscopy. *Skin Res. Technol.* **2020**, *26*, 956–959. [CrossRef] [PubMed]
23. Verzì, A.E.; Broggi, G.; Micali, G.; Sorci, F.; Caltabiano, R.; Lacarrubba, F. Line-Field Confocal Optical Coherence Tomography of Psoriasis, Eczema and Lichen Planus: A Case Series with Histopathological Correlation. *J. Eur. Acad. Dermatol. Venereol.* **2022**, *36*, 1884–1889. [CrossRef]
24. Ruini, C.; Schuh, S.; Hartmann, D.; French, L.; Welzel, J.; Sattler, E. Noninvasive Real-Time Imaging of Mite Skin Infestations with Line-Field Confocal Optical Coherence Tomography. *Br. J. Dermatol.* **2021**, *184*, e3. [CrossRef] [PubMed]
25. Ruini, C.; Schuh, S.; Pellacani, G.; French, L.; Welzel, J.; Sattler, E. In Vivo Imaging of Sarcoptes Scabiei Infestation Using Line-Field Confocal Optical Coherence Tomography. *J. Eur. Acad. Dermatol. Venereol.* **2020**, *34*, e808–e809. [CrossRef] [PubMed]
26. Ruini, C.; Schuh, S.; Gust, C.; Kendziora, B.; Frommherz, L.; French, L.E.; Hartmann, D.; Welzel, J.; Sattler, E. Line-Field Optical Coherence Tomography: In Vivo Diagnosis of Basal Cell Carcinoma Subtypes Compared with Histopathology. *Clin. Exp. Dermatol.* **2021**, *46*, 1471–1481. [CrossRef] [PubMed]
27. Suppa, M.; Fontaine, M.; Dejonckheere, G.; Cinotti, E.; Yélamos, O.; Diet, G.; Tognetti, L.; Miyamoto, M.; Orte Cano, C.; Perez-Anker, J.; et al. Line-Field Confocal Optical Coherence Tomography of Basal Cell Carcinoma: A Descriptive Study. *J. Eur. Acad. Dermatol. Venereol.* **2021**, *35*, 1099–1110. [CrossRef] [PubMed]
28. Verzì, A.E.; Micali, G.; Lacarrubba, F. Line-Field Confocal Optical Coherence Tomography May Enhance Monitoring of Superficial Basal Cell Carcinoma Treated with Imiquimod 5% Cream: A Pilot Study. *Cancers* **2021**, *13*, 4913. [CrossRef]
29. Ruini, C.; Schuh, S.; Gust, C.; Hartmann, D.; French, L.E.; Sattler, E.C.; Welzel, J. In-Vivo LC-OCT Evaluation of the Downward Proliferation Pattern of Keratinocytes in Actinic Keratosis in Comparison with Histology: First Impressions from a Pilot Study. *Cancers* **2021**, *13*, 2856. [CrossRef]
30. Dejonckheere, G.; Suppa, M.; Del Marmol, V.; Meyer, T.; Stockfleth, E. The Actinic Dysplasia Syndrome-Diagnostic Approaches Defining a New Concept in Field Carcinogenesis with Multiple CSCC. *J. Eur. Acad. Dermatol. Venereol.* **2019**, *33* (Suppl. S8), 16–20. [CrossRef]
31. Cinotti, E.; Tognetti, L.; Cartocci, A.; Lamberti, A.; Gherbassi, S.; Orte Cano, C.; Lenoir, C.; Dejonckheere, G.; Diet, G.; Fontaine, M.; et al. Line-Field Confocal Optical Coherence Tomography for Actinic Keratosis and Squamous Cell Carcinoma: A Descriptive Study. *Clin. Exp. Dermatol.* **2021**, *46*, 1530–1541. [CrossRef] [PubMed]
32. Ruini, C.; Schuh, S.; Gust, C.; Kendziora, B.; Frommherz, L.; French, L.E.; Hartmann, D.; Welzel, J.; Sattler, E.C. Line-Field Confocal Optical Coherence Tomography for the in Vivo Real-Time Diagnosis of Different Stages of Keratinocyte Skin Cancer: A Preliminary Study. *J. Eur. Acad. Dermatol. Venereol.* **2021**, *35*, 2388–2397. [CrossRef] [PubMed]
33. Lenoir, C.; Perez-Anker, J.; Diet, G.; Tognetti, L.; Cinotti, E.; Trépant, A.; Rubegni, P.; Puig, S.; Perrot, J.; Malvehy, J.; et al. Line-Field Confocal Optical Coherence Tomography of Benign Dermal Melanocytic Proliferations: A Case Series. *J. Eur. Acad. Dermatol. Venereol.* **2021**, *35*, e399–e401. [CrossRef]
34. Lenoir, C.; Cinotti, E.; Tognetti, L.; Orte Cano, C.; Diet, G.; Miyamoto, M.; Rocq, L.; Trépant, A.-L.; Perez-Anker, J.; Puig, S.; et al. Line-Field Confocal Optical Coherence Tomography of Actinic Keratosis: A Case Series. *J. Eur. Acad. Dermatol. Venereol.* **2021**, *35*, e900–e902. [CrossRef]
35. Lenoir, C.; Diet, G.; Cinotti, E.; Tognetti, L.; Orte Cano, C.; Rocq, L.; Trépant, A.-L.; Monnier, J.; Perez-Anker, J.; Rubegni, P.; et al. Line-Field Confocal Optical Coherence Tomography of Sebaceous Hyperplasia: A Case Series. *J. Eur. Acad. Dermatol. Venereol.* **2021**, *35*, e509–e511. [CrossRef] [PubMed]
36. Weedon, D. *Weedon's Skin Pathology*, 3rd ed.; Churchill Livingstone Elsevier: London, UK, 2010.
37. Gust, C.; Schuh, S.; Welzel, J.; Daxenberger, F.; Hartmann, D.; French, L.E.; Ruini, C.; Sattler, E.C. Line-Field Confocal Optical Coherence Tomography Increases the Diagnostic Accuracy and Confidence for Basal Cell Carcinoma in Equivocal Lesions: A Prospective Study. *Cancers* **2022**, *14*, 1082. [CrossRef] [PubMed]


**Disclaimer/Publisher's Note:** The statements, opinions and data contained in all publications are solely those of the individual author(s) and contributor(s) and not of MDPI and/or the editor(s). MDPI and/or the editor(s) disclaim responsibility for any injury to people or property resulting from any ideas, methods, instructions or products referred to in the content.

diagnostics

*Article*

# VIS-NIR Diffuse Reflectance Spectroscopy System with Self-Calibrating Fiber-Optic Probe: Study of Perturbation Resistance

**Valeriya Perekatova, Alexey Kostyuk, Mikhail Kirillin *, Ekaterina Sergeeva, Daria Kurakina, Olga Shemagina, Anna Orlova, Aleksandr Khilov and Ilya Turchin**

Institute of Applied Physics RAS, 603950 Nizhny Novgorod, Russia
* Correspondence: mkirillin@yandex.ru

**Abstract:** We report on the comparative analysis of self-calibrating and single-slope diffuse reflectance spectroscopy in resistance to different measurement perturbations. We developed an experimental setup for diffuse reflectance spectroscopy (DRS) in a wide VIS-NIR range with a fiber-optic probe equipped with two source and two detection fibers capable of providing measurements employing both single- and dual-slope (self-calibrating) approaches. In order to fit the dynamic range of a spectrometer in the wavelength range of 460–1030 nm, different exposure times have been applied for short (2 mm) and long (4 mm) source-detector distances. The stability of the self-calibrating and traditional single-slope approaches to instrumental perturbations were compared in phantom and in vivo studies on human palm, including attenuations in individual channels, fiber curving, and introducing optical inhomogeneities in the probe–tissue interface. The self-calibrating approach demonstrated high resistance to instrumental perturbations introduced in the source and detection channels, while the single-slope approach showed resistance only to perturbations introduced into the source channels.

**Keywords:** diffuse reflectance spectroscopy; diffuse optical spectroscopy; tissue optics; diffuse scattering; oxygenation; tissue chromophores; self-calibrating approach; ratiometric approach

**Citation:** Perekatova, V.; Kostyuk, A.; Kirillin, M.; Sergeeva, E.; Kurakina, D.; Shemagina, O.; Orlova, A.; Khilov, A.; Turchin, I. VIS-NIR Diffuse Reflectance Spectroscopy System with Self-Calibrating Fiber-Optic Probe: Study of Perturbation Resistance. *Diagnostics* **2023**, *13*, 457. https://doi.org/10.3390/diagnostics13030457

Academic Editor: Viktor Dremin

Received: 15 December 2022
Revised: 24 January 2023
Accepted: 24 January 2023
Published: 26 January 2023

**Copyright:** © 2023 by the authors. Licensee MDPI, Basel, Switzerland. This article is an open access article distributed under the terms and conditions of the Creative Commons Attribution (CC BY) license (https://creativecommons.org/licenses/by/4.0/).

## 1. Introduction

Diffuse reflectance spectroscopy (DRS) is an optical technique that allows the evaluation of tissue biochemistry and microstructure for a number of applications including brain hemodynamics [1] also called fNIRS, diagnostics of breast tumor margins [2,3] and treatment monitoring [4], skin cancer diagnostics [5–7], evaluating the scar severity and therapeutic response of keloid [8], and diagnostics of tumor margins in the oral cavity (head and neck cancer) [9], lung [10], liver [11–13], and colon [14,15]. A number of potential applications have also been reported, such as diagnostics of thyroid [16] and adipose tissue [17] and the identification of neurovascular bundles. The DRS principle is based on delivering broadband light to the biotissue and registering the backscattered light at a specified distance. The detected signal contains information about scattering (related to the microstructure of the tissue) and absorption (related to its biomolecular composition). Due to the strong dependence of absorption coefficients of different chromophores (oxy- and deoxyhemoglobin, melanin, lipids, water, etc.) on the wavelength, one can reconstruct their concentrations in tissue by analyzing the extinction of the light spectrum between the source and detector.

The DRS probing spectral range is selected depending on the absorption spectra of the studied chromophores and the desired probing volume in tissue. For example, the concentrations of oxy- and deoxyhemoglobin in superficial tissues can be reconstructed using the visible (usually 500–600 nm) spectral range, while for deeper probing it is reasonable to use the range of 700–900 nm due to higher light penetration depth. The NIR range is also used to assess water and lipid content [18], while the wider VIS-NIR range can be used for analysis of collagen and elastin content [8]. Currently, VIS-NIR spectroscopy has been applied in several works and has shown higher potential in comparison with VIS or NIR spectroscopy separately [14,19,20], because it allows the reconstruction of concentrations of a larger set of tissue chromophores and/or obtaining a higher accuracy [18,21].

If tissue optical properties vary with depth, DRS in the VIS-NIR range can be applied to assess chromophore concentrations in different tissue layers using differences in sensitivity depths of the VIS and NIR spectrum regions. This approach was successfully applied to assess skin hemoglobin concentrations in the dermis and lower dermis [22].

The results of the reconstruction of tissue chromophores in DRS have the following keys to success: (1) the applied light transport model should be realistic enough to correctly describe light attenuation from source to detector; (2) the reconstruction procedure should have a good convergence; (3) instrumental characteristics, such as source spectral brightness, detector spectral sensitivity, transient characteristics of source and detector fiber channels, and optical contact between the DRS probe and tissue should be taken into account. In the present study we concentrate mainly on the last issue because incorrect consideration of the instrumental characteristics can lead to significant errors in the reconstructed values even with an appropriate light transport model and a valid reconstruction technique. This aspect is especially essential in broadband measurements, in particular VIS-NIR, due to possible light dispersion in the instrumental part and strong differences in light attenuation in tissue in different spectral ranges, and, therefore, a need for adjusted compensation.

Different strategies are applied in DRS to correctly account for the instrumental characteristics depending on the measured data type. For example, investigation of hemodynamics based on the measurement of relative changes in hemoglobin concentrations in time can be implemented using a simple single source-detector distance (SDD) approach with a single source-detector pair. The resulting equation for relative changes in hemoglobin concentrations in time allows for the exclusion of instrumental characteristics [23]. Absolute measurements of chromophore concentrations in a single SDD configuration are usually accompanied by calibration measurements with a tissue phantom with known absorption and scattering characteristics or a reflection standard [24]. However, if instrumental characteristics vary in time (for example, source spectral brightness may significantly vary in lamp sources), a calibration procedure should be applied periodically, which is not convenient or even impossible during continuous biomedical examination. Continuous calibration measurements can be provided with the help of an additional source-detector channel with a reflection standard at the tips of source and detection optical fibers [25].

Two SDDs with a single source and two detectors or a single detector and two sources allow the assessment of effective light attenuation $\mu_{eff}$ in tissue by taking a ratio between the detected signals obtained at different SDDs. In this ratiometric approach, also called single-slope measurement, most of the instrumental characteristics are excluded in the final equation for $\mu_{eff}$, which yields a more accurate assessment of tissue chromophore content in comparison with the single-distance approach [26]. Multiple sources at a single detector or multiple detectors at a single source are used to increase the precision of the extinction coefficient extraction. However, the effect of instrumental function is not completely eliminated in this approach.

A possible solution to compensate for more instrumental contributions is a self-calibrating technique suggested in [27]. The idea is based on symmetrical multi-distance measurements; at least four measurements at each wavelength with two sources and two detectors with a symmetrical configuration (Figure 1) should be provided to obtain calibration-free characterization of the studied tissue. This probe demonstrated more reliable data on the optical properties of tissue and higher long-term stability compared to standard DRS configuration due to a reduction in instrumental errors.

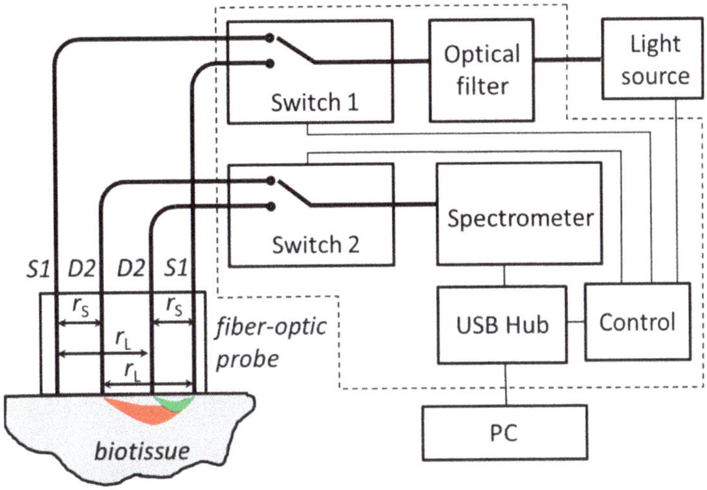


**Figure 1.** Schematic of experimental setup with self-calibrating fiber-optic probe. S1,2: the source fibers; D1,2: the detection fibers; areas of effective sensitivities for short ($r_S$) and long ($r_L$) SDDs are marked with green and red colors, respectively. Bold lines indicate optical fibers, thin lines indicate electrical connections. Elements within the dashed frame are placed in a single housing.

In addition to insensitivity toward instrumental effects, the self-calibrating approach is less sensitive to the changes in optical coupling between the optical probe and tissue. The last advantage is tightly connected with the differences in sensitivity to superficial or deeper chromophores in different approaches: the traditional single-measurement approach has a banana-shaped sensitivity function [28] with maxima near source– and detector–tissue interfaces, while the self-calibrating technique is relatively more sensitive to deeper tissues [29,30]. This finding makes the self-calibrating approach very attractive in its application for studies of brain activity in the NIR spectral region [31]. Single-distance or single-slope approaches register primarily photons backscattered from the scalp and skull masking the brain hemodynamics, and the traditional increase in SDD does not provide any significant benefit, since the maximum sensitivity remains near the source– and detector–tissue contacts. The self-calibrating approach was applied for diagnostics of breast tumors with a more sophisticated probe including 16 continuous-wave (CW) sources at 690 nm and 830 nm and 8 detectors located symmetrically [32]. Multiple sources and detectors allow obtaining a signal averaged over a large tissue volume resulting in more robust data on the oxygenation of tumor tissue [33].

The main drawback of all pure CW optical diffuse measurements is related to the difficulty of separating absorption $\mu_a$ and reduced scattering $\mu_s'$ coefficients which are included in the expression for the effective extinction coefficient of diffuse light as a product:

$$\mu_{eff} = \sqrt{3\mu_a(\mu_a + \mu_s')}. \tag{1}$$

Employing the reduced scattering coefficient values from the literature may cause significant errors in absolute measurements of chromophore concentrations due to tissue-to-tissue variations in reduced scattering values [34]. Additional frequency-domain (FD) measurements employing high-frequency modulation of probing light intensity at two or more wavelengths allow assessing reduced scattering directly at these wavelengths. Assuming a power law or a linear wavelength dependence of reduced scattering coefficient (the latter approach is reasonable in NIR where the decrease in $\mu_s'(\lambda)$ dependence is slow), one can estimate it within the whole spectrum measured by CW DRS, which results in higher precision of chromophores reconstruction [35] (steady-state and frequency-domain

(SSFD) reflectance measurements). SSFD measurements at two wavelengths were also successfully applied in combination with a self-calibrating approach for studying human brain hemodynamics [36,37].

The FD technique is applied successfully for the NIR spectral range for large SDDs that ensure a sufficient phase shift of a photon density wave propagated from source to detector. However, at large SDDs (approx. more than 5 mm), VIS measurements are restricted by high attenuation of light in biotissue. Smaller SDDs (<5 mm) need higher modulation frequency to register the phase shift that is hard to implement technically. In this connection, DRS in VIS range 500–600 nm is usually performed with the CW technique alone [26]. In the joint VIS-NIR range, the reduced scattering spectrum $\mu_s'(\lambda)$ has a more sophisticated behavior and can be approximated by a sum of Rayleigh-scattering and Mie-scattering components [18], since Rayleigh scattering is assumed to dominate in the UV-blue optical range, while Mie scattering prevails in NIR:

$$\mu_s'(\lambda) = a\left[f\left(\frac{\lambda}{\lambda_0}\right)^{-4} + (1-f)\left(\frac{\lambda}{\lambda_0}\right)^{-b}\right] \tag{2}$$

Here the parameter $a$ is the reduced scattering coefficient at $\lambda_0$ = 500 nm, $f$ is the fraction of Rayleigh scattering with $\lambda^{-4}$ dependence that is described by the first term in the brackets, and $b$ is the power index of Mie scattering wavelength dependence described by the second term. Parameters $a$, $b$, and $f$ can be assessed along with unknown tissue chromophores composition from the fitting of an experimentally obtained reflected spectrum by a model function. This approach has been applied in VIS-NIR DRS for a simplified $\mu_s'(\lambda)$ dependence [38].

In this paper, we present a comparative analysis of the sensitivity of the dual- and single-slope approaches in DRS to various perturbations than can occur in the course of measurements. The analysis is based on the assessment of the changes in the reconstructed effective extinction coefficient spectrum in response to the introduced instrumental distortion. The study is performed using a custom-built experimental setup for VIS-NIR DRS with a fiber-optic probe employing a self-calibrating approach. To the best of our knowledge, this is the first application of a self-calibrating approach for ultra-wideband 460–1030 nm (VIS-NIR) DRS. The problem that arises is the significant difference in absorption coefficients in VIS and NIR spectral regions, leading to difficulties in detecting both regions simultaneously with a single spectrometer at different SDDs with a large enough signal-to-noise ratio. In the proposed experimental setup, we solved this problem by applying individual exposure times for small and large SDDs in order to fit the whole received spectra for both SDDs in the spectrometer dynamic range. The reconstruction of tissue optical parameters is proposed via a minimization in the difference between the effective extinction coefficient $\mu_{eff}$, which is evaluated from DRS measurements using simplified light diffusion theory, and the expected model coefficient $\mu_{eff}$ calculated with Equation (1). The absorption coefficient in Equation (1) is assumed to be a linear combination of basic biotissue chromophores absorption spectra, while the reduced scattering coefficient is described by Equation (2). The developed experimental setup has been tested for resistance toward different instrumental perturbations including the bending of optical fibers, installing an additional attenuator in an individual channel, and modifying the probe–tissue interface. The results for the self-calibrating approach were compared to those for the single-slope measurements.

## 2. Materials and Methods

### 2.1. Evaluation of Extinction Coefficient with Self-Calibrating Approach Measurements

Propagation of light between source and detector in biological tissue is well described by the Radiative Transfer Equation (RTE) [39], employing absorption $\mu_a$ and scattering $\mu_s$ indices and a scattering phase function as tissue optical properties. There exists no general analytical solution to this equation, however, for a number of applications with SDD exceeding several light transport lengths, the RTE can be reduced to the diffuse equation, which has an analytical solution for homogeneously scattering and absorbing media. In the frame of this approach, a photon fluence rate generated by a point light source with unit power in the infinite medium (Green's function) is defined by equation [40]:

$$\phi(r) = \frac{3(\mu_a + \mu_s')}{4\pi r}\exp\left(-\mu_{eff}r\right) \quad (3)$$

where $r$ is the distance from the source and $\mu_{eff}$ is defined by Equation (1) via the absorption coefficient $\mu_a$ and reduced scattering coefficient $\mu_s'$. Under certain assumptions (neglecting medium boundary conditions and radiating patterns of source and detection fibers), Equation (3) can be used to characterize light intensity in the detection fiber $Dj$ located at distance $r_{kj}$ from the source fiber $Sk$ ($j$,$k$ = 1,2) in the configuration shown in Figure 1. Following these assumptions, the DRS signal registered by a spectrometer at the specific wavelength can be written in the following form:

$$I_{kj} = \frac{3(\mu_a + \mu_s')AS_kAD_jI_0\eta}{4\pi r_{kj}}E_{kj}\exp\left(-\mu_{eff}r_{kj}\right) \quad (4)$$

where $I_0$ is a light source intensity; $AS_k$ and $AD_j$ are the transient characteristics of DRS device parts describing the propagation of light from the light source to a contact of the source fiber $k$ with the biotissue and from the biotissue contact with the detection fiber $j$ to the spectrometer, respectively; $\eta$ is the spectrometer sensitivity; and $E_{kj}$ is the corresponding spectrometer exposure time.

The ratio of signals received with a common source $k$ and two detectors $j$ = 1,2 having different SDDs (known as single-slope configuration) carries information about the medium extinction coefficient and excludes most of instrumental characteristics (source transient characteristics, source brightness, and spectrometer sensitivity):

$$\frac{I_{k1}}{I_{k2}} = \frac{AD_1r_{k2}E_{k1}}{AD_2r_{k1}E_{k2}}\exp\left(-\mu_{eff}(r_{k1} - r_{k2})\right) \quad (5)$$

Typically an assumption is made that both detection channels have equal transient characteristics, i.e., $AD_1 = AD_2$. In this case, one can evaluate the extinction coefficient following Equation (5) and using the ratio $I_{11}/I_{12}$ of the spectra measured by the system depicted in Figure 1 with source S1 and two detectors D1 and D2 (abbreviated as S1D1D2) as:

$$\mu_{eff} = \frac{1}{r_{12} - r_{11}}ln\left[\left(\frac{I_{11}}{I_{12}}\right)\left(\frac{E_{12}}{E_{11}}\right)\frac{r_{11}}{r_{12}}\right] \quad (6)$$

and using the ratio $I_{22}/I_{21}$ of the spectra measured with source S2 and two detectors D1 and D2 (abbreviated as S2D1D2) as:

$$\mu_{eff} = \frac{1}{r_{21} - r_{22}}ln\left[\left(\frac{I_{22}}{I_{21}}\right)\left(\frac{E_{21}}{E_{22}}\right)\frac{r_{22}}{r_{21}}\right] \quad (7)$$

However, if the assumption $AD_1 = AD_2$ is incorrect, the extinction spectrum calculated by Equations (6) or (7) contains an error. In this case, for symmetrical measurements when $r_{11} = r_{22} = r_S$ and $r_{21} = r_{12} = r_L$, Equation (5) yields:

$$\begin{cases} \frac{I_{11}}{I_{12}} = \frac{AD_1r_LE_{11}}{AD_2r_SE_{12}}\exp\left(\mu_{eff}(r_L - r_S)\right) \\ \frac{I_{21}}{I_{22}} = \frac{AD_1r_SE_{21}}{AD_2r_LE_{22}}\exp\left(-\mu_{eff}(r_L - r_S)\right) \end{cases} \quad (8)$$

and the effective extinction coefficient can be derived in the form that excludes both source and detector transient characteristics:

$$\mu_{eff} = \frac{1}{2(r_L - r_S)} ln\left[\left(\frac{I_{11}I_{22}}{I_{12}I_{21}}\right)\left(\frac{E_{12}E_{21}}{E_{11}E_{22}}\right)\frac{r_S^2}{r_L^2}\right] \tag{9}$$

Equation (9) represents a self-calibrating (or dual-slope) approach, employing four source-detector measurements: S1-D1, S1-D2, S2-D1, S2-D2 (abbreviated as S1D1S2D2). One can note that expression (9) is an average value of the extinction coefficients obtained by Equations (6) and (7).

### 2.2. Reconstruction of Skin Chromophores and Scattering from $\mu_{eff}$ Spectrum

In this study we focus on the characterization of human skin scattering and absorption spectra. Multiplicative combinations of absorption coefficients and reduced scattering coefficients in expression (1) for the extinction coefficient makes their separation from the continuous wave measurements difficult, in contrast to frequency-domain or time-domain techniques. In our study, in order to separate $\mu_a$ from $\mu_s'$ based on Equation (1) and the extinction coefficient spectrum experimentally obtained with Equations (6), (7), or (9), depending on the employed measurement scheme, we make several empirical assumptions.

The absorption coefficient $\mu_a$ is considered as a weighted sum of the absorption spectra of basic skin chromophores. In our consideration we limit their set to melanin, blood, water, and "dry matter", assuming that the contribution of other chromophores in the range of 460–1030 nm is negligible:

$$\begin{gathered} \mu_a(\lambda) = C_{water} * \mu_a^{water}(\lambda) + C_{mel} * \mu_a^{mel}(\lambda) + \\ \left[C_{blood} * \left(StO_2 * \mu_a^{oxy}(\lambda) + (1 - StO_2) * \mu_a^{deoxy}(\lambda)\right)\right] + \mu_a^{dry} \end{gathered} \tag{10}$$

where the spectra of $\mu_a^{HbO_2}$, $\mu_a^{Hb}$, and $\mu_a^{mel}$ are taken from [41,42]. Absorption of "dry matter" is taken as a wavelength independent $\mu_a^{dry}$. A reduced scattering coefficient spectrum is taken according to Equation (2) with the Rayleigh fraction which approximates scattering at small wavelengths and Mie fraction which dominates in NIR. The contribution of lipids to the absorption coefficient is not considered in this study since the SDD in the described system is limited by 4 mm and the measurements described below have been performed on human palm with low lipid content. The parameters of oxygen saturation $StO_2$ concentration of different chromophores,—$C_{water}$, $C_{mel}$, $C_{blood}$, $\mu_a^{dry}$—together with the scattering parameters $a$, $b$, and $f$ from Equation (2), can be determined by fitting the experimentally obtained extinction spectrum $\mu_{eff}$ with a combination of empirical dependencies (10) and (2) substituted into Equation (1). This optimization problem was solved numerically by finding the parameters vector $K = [\,C_{blood}, StO_2, C_{water}, \mu_a^{dry}, a, b, f]$ using lsqcurvefit MATLAB function within the wavelength range of 460–1030 nm assuming $C_{mel}$ to be a constant equal to 0.005. This solution can be derived using Equation (9) in the case of the self-calibrating approach as:

$$K = \underset{K}{\operatorname{argmin}} \sum_{\lambda} \left(\sqrt{3\mu_a(K)(\mu_a(K) + \mu_s'(K))} - \frac{1}{2(r_L - r_S)} ln\left[\left(\frac{I_{11}I_{22}}{I_{12}I_{21}}\right)\left(\frac{E_{12}E_{21}}{E_{11}E_{22}}\right)\frac{r_S^2}{r_L^2}\right]\right)^2 \tag{11}$$

or in the case of single-slope approach in S1D1D2 mode using Equation (6) as:

$$K = \underset{K}{\operatorname{argmin}} \sum\nolimits_{\lambda} \left(\sqrt{3\mu_a(K)(\mu_a(K) + \mu_s'(K))} - \frac{1}{r_L - r_S} ln\left[\left(\frac{I_{11}}{I_{12}}\right)\left(\frac{E_{12}}{E_{11}}\right)\frac{r_S}{r_L}\right]\right)^2, \tag{12}$$

or in S2D1D2 mode using Equation (7) as:

$$K = \underset{K}{\operatorname{argmin}} \sum_{\lambda} \left( \sqrt{3\mu_a(K)(\mu_a(K) + \mu_s'(K))} - \frac{1}{r_L - r_S} ln \left[ \left( \frac{I_{22}}{I_{21}} \right) \left( \frac{E_{21}}{E_{22}} \right) \frac{r_S}{r_L} \right] \right)^2. \quad (13)$$

During the optimization procedure the following limitations were set on the extracted variables: $StO_2$ varies in the physiological range of [0, 1], $a$ varies in the range of [0.5, 10] $mm^{-1}$, $b$ varies in the range of [0, 3], $f$ varies in the range of [0, 1], volume fractions $C_{blood}$ and $C_{water}$ vary within [0, 1], and $\mu_a^{dry}$ varies within [0, ∞]. Lower and upper limits for the parameters $a$ and $b$ are determined according to the reported data on the range of the skin reduced scattering coefficient at $\lambda_0 = 500$ nm and the corresponding power index values [18,43]. Since $\mu_a^{mel}(\lambda)$ and $\mu_s'(\lambda)$ both monotonously decrease with the wavelength, high uncertainty arises in the joint reconstruction of the parameters of $C_{mel}$, $a$, $b$, and $f$; therefore, $C_{mel}$ was chosen as a constant value.

### 2.3. Experimental DRS Setup

An experimental DRS setup with a self-calibrating fiber-optic probe was constructed in accordance with the scheme shown in Figure 1. Radiation from Fiber-Coupled Xenon (SLS205, Thorlabs, Newton, NJ, USA) broadband 240–1200 nm light source was used for tissue probing. The source has a mechanical shutter driven by a TTL pulse from the Control unit Arduino Uno (Arduino, Scarmagno, Italy). It has rather high spectral radiance both in VIS and NIR spectral regions in comparison with those traditionally applied in VIS-NIR spectroscopy tungsten halogen lamps which have low spectral radiance at wavelengths below 600 nm. Probing light passes through BS-8 (Zapad Pribor, Moscow, Russia) absorption filter cutting UV light below 380 nm placed in the fiber-optic FOFMS/M (Thorlabs Inc., Newton, NJ, USA) filter holder. A UV filter is used to prevent possible negative effects on biological tissue and solarization of the probe and switch optical fibers. Spectrally corrected light passes through 1 × 2 fiber-optical switch 1 (Piezosystem Jena GmbH, Jena, Germany) which selects the $S1$ or $S2$ source fiber of the probe to illuminate biological tissue. Diffusively scattered light is detected by detection fibers $D1$ and $D2$, one of which is selected by a 1 × 4 switch 2 (Piezosystem Jena GmbH, Jena, Germany) to deliver light to spectrometer Maya 2000 PRO (Ocean Optics, Orlando, FL, USA). Switch 2 uses only two outputs for the applied optical probe which has only two detection fibers. Both switches are driven by the control unit. The Maya spectrometer has rather high sensitivity in NIR up to 1100 nm which covers water and lipid absorption bands near 980 nm and has high linearity due to the applied calibration which is essential for the application of a self-calibrating approach. The spectrometer exposure times are set individually for small and large SDDs to fit the spectrometer dynamic range. In the current in vivo and phantom studies, the exposure times were set equal to $E_{21} = E_{12} = 80$ ms for long SDD $r_L$= 4 mm and $E_{11} = E_{22} = 15$ ms for short SDD $r_S$=2 mm. For each source–detector pair, the detected signal is averaged over several subsequent measurements in order to increase signal-to-noise ratio with the following subtraction of a dark signal obtained at the closed source shutter with the same exposure time and averaging. For the in vivo studies, the averaging number is taken from the ratio $T/E_{21}$ or $T/E_{11}$ where T is a heartbeat period. The resulting spectra $I_{11}$, $I_{12}$, $I_{21}$, and $I_{22}$ are stored at PC for further analysis described below in Section 2.3.

Domestically designed probe housing is made of black photopolymer resin Anycubic Basic (HONGKONG Anycubic Technology Co., Limited, Hong Kong, China) by 3D printing on a Phrozen Shuffle 2019 (Hsinchu City, Taiwan) printer and has 6 × 8 $mm^2$ area of the probe–tissue interface. Two 400 µm source fibers and two 200 µm detection fibers with 0.22 NA (Thorlabs, Newton, NJ, USA) were placed in line inside the probe housing at a 2 mm distance between neighboring fibers, which results in short and long SDDs of $r_S$ = 2 mm and $r_L$ = 4 mm, respectively.

The fiber-optic probe is equipped with a mechanical pressure control unit. The pressure was set equal to 12.7 kPa in all studies.

The DRS system is fully automated by a JAVA code operated with a source shutter, spectrometer, and fiber switches with the help of a control unit. Full acquisition time was about 6 s for the abovementioned exposure times and a heartbeat period of 1 s.

### 2.4. Instrumental Perturbations in Phantom and In Vivo DRS Studies

The developed experimental setup was tested on a silicone-based biotissue phantom employed as a reference standard for DRS measurements in [26] and on a palm of a healthy volunteer from the group of the researchers. In both series of experiments, various types of instrumental perturbations were applied to a developed DRS setup in order to compare the stability of the self-calibrating and single-slope approaches, including installing an additional attenuator in an individual channel, bending the optical fiber, and modifying the probe–tissue interface (Table 1).

**Table 1.** Description of instrumental perturbations used in phantom and in vivo experiments.

| Perturbation Index *P* | Description of Perturbation |
|---|---|
| D1L | Simulation of losses in detection channel. Ø105 μm fiber was inserted between detection fiber D1 and Switch 2. |
| D1C | Simulation of curving detection fiber. Detection fiber D1 is curved into a 50 mm radius ring. |
| D1B | Simulation of probe–tissue interface inhomogeneity in detection channel. A blue sticker was inserted between detection fiber D1 end and investigated object |
| D1G | Simulation of probe–tissue interface inhomogeneity in detection channel. A green sticker was inserted between detection fiber D1 end and investigated object |
| D1P | Simulation of probe–tissue interface inhomogeneity in detection channel. A pink sticker was inserted between detection fiber D1 end and investigated object |
| S1L | Simulation of losses in source channel. Ø105 μm fiber was inserted between source fiber S1 and Switch 1. |
| S1C | Simulation of bending source fiber. Source fiber S1 is curved into 50 mm radius ring. |
| S1B | Simulation of probe–tissue interface inhomogeneity in source channel. A blue sticker was inserted between source fiber S1 end and investigated object |
| S1G | Simulation of probe–tissue interface inhomogeneity in source channel. A green sticker was inserted between source fiber S1 end and investigated object |
| S1P | Simulation of probe–tissue interface inhomogeneity in source channel. A pink sticker was inserted between source fiber S1 end and investigated object |
| INIT | No perturbations were implemented to fibers |

Installing additional fibers (labeled as D1L and S1L in Table 1) with a smaller core diameter of 105 μm (compared to 200 and 400 μm fibers used in the setup) in an individual channel simulates possible losses in fiber-optic contacts between different instrumental parts: the light source, detector, optical switches, and fiber-optic probe. Curving a probe fiber (labeled as D1C and S1C in Table 1) into a 50 mm radius ring simulates the fiber curvature occurring in the course of a medical procedure when a fiber-optic probe examines different tissue localizations and fibers are randomly bent. Modification of the probe–tissue interface with plastic page stickers (series NEON, BRAUBERG, Frankfurt, Germany) with different colors (blue, green, and pink labeled as D1B and S1B, D1" and S1G, and D1P and S1P, respectively) simulates random biotissue surface inhomogeneities that are always present in biological tissue examinations.

In order to quantify the spectral effects of all the studied perturbations, we measured the spectral transfer functions of the introduced perturbation (the ratio of the measured spectrum with the perturbation introduced to that in the absence of the perturbation). The measurement results are shown in Figure 2 and demonstrate that most of the considered perturbations lead to spectrum shape distortion which may potentially lead to errors in reconstruction of physiological properties from the DRS measurements.

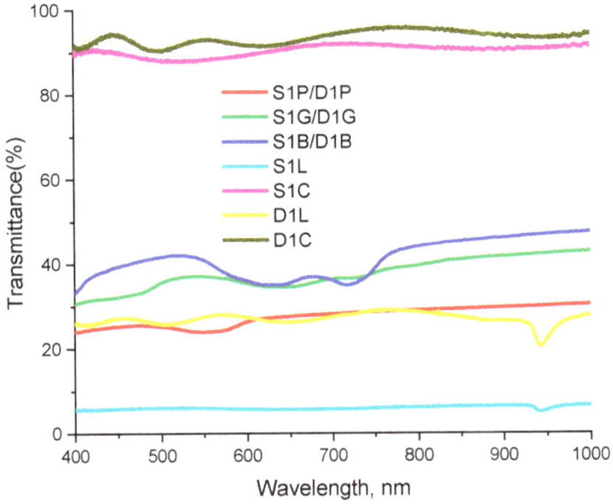


**Figure 2.** Transfer functions of the studied perturbations (the ratio of the measured spectrum with the perturbation introduced to that in the absence of the perturbation). Perturbations are indicated in accordance with Table 1.

The colored stickers naturally feature transmission bands corresponding to their visible colors. Note a significant difference in the transmittance coefficient for perturbations S1L and D1L consisting in the insertion of an additional fiber to the source or the detection channel, respectively, that originates from the different mismatch between the diameters of the original fibers and the inserted fiber. For each perturbation, DRS measurements were repeated 3 times; before each measurement the position of a probe was slightly changed by removing and then replacing the probe. Unperturbed measurements were repeated 6 times: 3 times before applying perturbations and 3 times after.

The perturbations applied to source S2 and detector D2 are not listed in Table 1 because they provide similar results to the perturbations applied to S1 and D1 for reasons of symmetry (Figure 1).

### 2.5. Calculation of Extinction Spectra Variations

Various instrumental perturbations listed in Table 1 result in different deviations in the extinction spectrum calculated by Equations (6), (7), or (9). These deviations are quantified as a root mean square deviation (RMSE) of the extinction spectrum values $\mu_{eff}^{P,m}(\lambda)$ evaluated under particular perturbation in a single experiment from the initial $\mu_{eff}^{INIT}$ extinction spectrum measured without perturbations:

$$\Delta\mu_{eff}^{P,m} = \sqrt{\frac{\sum_{\mathbf{i=1}}^{\mathbf{N}_\lambda}\left(\mu_{eff}^{P,m}(\lambda_i) - \mu_{eff}^{INIT}(\lambda_i)\right)^2}{\mathbf{N}_\lambda}} \quad (14)$$

where $\lambda_i$ is the $i$-th wavelength, $i = 1 \ldots N$; $P$ is the perturbation index listed in Table 1, and $m$ is the measurement number with the particular perturbation $m = 1 \ldots N_P$. For perturbed measurements, $N_{P \neq INIT} = 3$, and for unperturbed measurements, $N_{P=INIT} = 6$. The initial extinction spectrum $\mu_{eff}^{INIT}(\lambda_i)$ is calculated as an average value for each wavelength over

6 measurements provided without any perturbations. To quantify deviations caused by perturbations of type $P$, an average value over $\Delta\mu_{eff}^{P,m}$ is calculated:

$$\Delta\mu_{eff}^{P} = \frac{\sum_{m=1}^{N_P} \Delta\mu_{eff}^{P,m}}{N_P} \quad (15)$$

Variations of reconstructed tissue chromophore concentrations and scattering properties obtained for in vivo measurements were calculated in the same way.

## 3. Results

### 3.1. Phantom DRS Measurements

Several DRS phantom measurements have been performed with and without the instrumental perturbations indicated in Table 1.

Figure 3 shows the examples of extinction spectra $\mu_{eff}^{P,m}$ of a silicone phantom calculated using the single-slope approach in S1D1D2 and S2D1D2 configurations and the calibration-free approach in S1S2D1D2 configuration for different kinds of source (S1L, S1C, S1B) and detector (S1L, S1C, S1B) perturbations, as well as for unperturbed data. As one can see from Figure 3a,b,f, all extinction spectrum curves calculated using the self-calibrating approach are close to each other, which indicates a high resistance to instrumental perturbations introduced to source or detector channels (Figure 3a,b). In contrast, the single-slope approach demonstrates resistance only to source perturbations (Figure 3(d)), while perturbations introduced to the detector channel lead to significant variations in $\mu_{eff}$ (Figure 3c,e). It can be seen from Figure 3c,e that if a perturbation $P$ results in an increase in $\mu_{eff}^{P}$ values calculated by S1D1D2 data, the value of $\mu_{eff}^{P,m}$ calculated by S2D1D2 data decreases. For example, the absorption band of the blue sticker (see Figure 2) employed in perturbation D1B manifests by the deformation of the $\mu_{eff}$ spectrum reconstructed by single-slope measurements in S1D1D2 and S2D1D2 configurations in opposite ways (Figure 3(c) and 3(e), respectively) according to Equations (6) and (7). The introduction of loss perturbation D1L appears as a negative (Figure 3c) or positive (Figure 3e) shift in the reconstructed $\mu_{eff}$ spectrum together with a variation around 950 nm which is determined by the transmission peak in the transfer function of this perturbation (Figure 2). The fiber curving perturbation D1C provides minimal variations in $\mu_{eff}$ spectrum since it has a transmittance close to 100% (Figure 2). The increase in the noise level in Figure 3b,d for S1L perturbation is caused by a drop in light intensity, while the shape of the extinction spectra does not change in both S1D1D2 and S1S2D1D2 cases.

Figure 3f demonstrates variations in the extinction spectra of the uniform silicone phantom for repeated unperturbed measurements. Variations for the single-slope configurations are higher than those for the self-calibrating approach. It should also be noted that red and blue curves corresponding to $\mu_{eff}$ calculated by the single-slope approach in the two configurations S1D1D2 and S2D1D2, respectively, are positioned above and below the gray curve for $\mu_{eff}$ recovered by the self-calibrating approach, which is associated with non-identical transient characteristics of the detection channels D1 and D2.

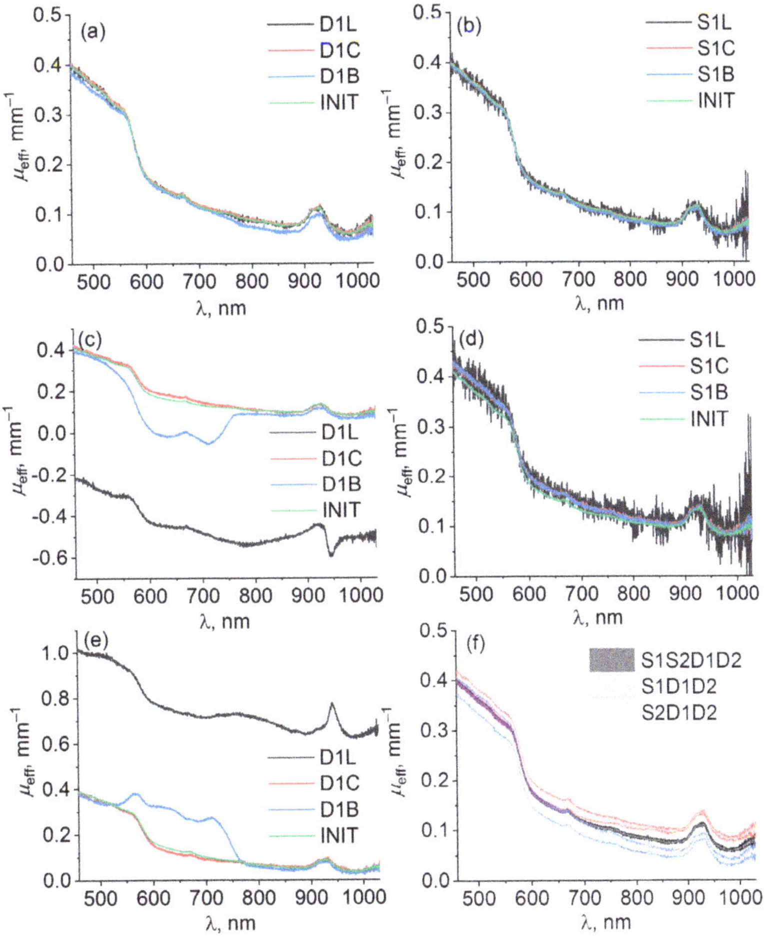


**Figure 3.** Spectra of $\mu_{eff}^{P,m}$ of a biotissue phantom calculated with self-calibrating approach S1S2D1D2 (**a**,**b**) and single-slope approach in configurations S1D1D2 (**c**,**d**) and S2D1D2 (**e**) for different kinds of detector (left column) and source (right column) perturbations listed in Table 1. Index m indicates individual measurement under particular perturbation P. Comparison of $\mu_{eff}^{INIT}$ spectra (**f**) obtained by three approaches S1S2D1D2 (gray), S1D1D2 (red), and S2D1D2 (blue) for unperturbed measurements shown as mean with confidence bounds.

### 3.2. In Vivo DRS Measurements of Human Skin

In vivo DRS measurements demonstrate almost similar results to the phantom studies (Figure 4). High resistance of the self-calibrating approach to both source (Figure 4b) and detector (Figure 4a) instrumental perturbations is observed for this case, whereas the single-slope approach has resistance only to source perturbations (Figure 4d), while detector perturbations can lead to a significant corruption of the reconstructed extinction spectrum (Figure 4c,e) such as those observed for the silicone phantom measurements (Figure 3c,e).

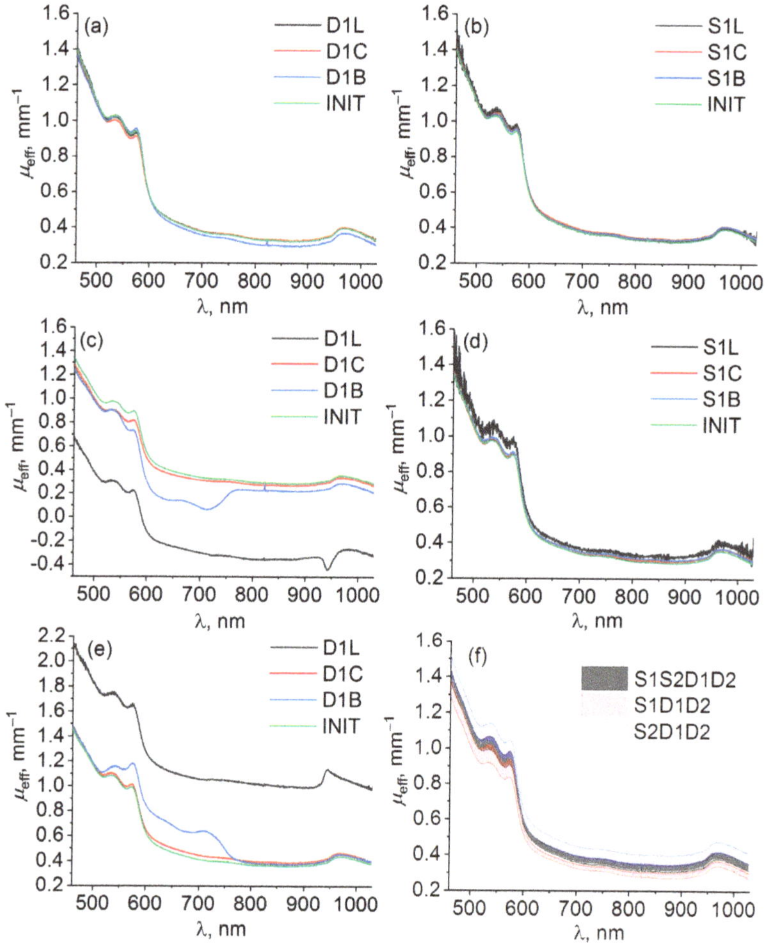


**Figure 4.** Examples of $\mu_{eff}^{P,m}$ spectra of a human palm calculated with self-calibrating approach S1S2D1D2 (**a**,**b**), and single-slope configurations S1D1D2 (**c**,**d**), and S2D1D2 (**e**) for different kinds of detector (left column) and source (right column) perturbations described in Table 1. Index m indicates individual measurements under particular perturbation P. Comparison of $\mu_{eff}^{INIT}$ spectra (**f**) obtained by three approaches S1S2D1D2 (gray), S1D1D2 (red), and S2D1D2 (blue) for unperturbed measurements shown as mean with confidence bounds.

In contrast to the results of phantom studies, the extinction coefficient in in vivo studies calculated for unperturbed (INIT) measurements (Figure 4f) demonstrate similar variations in self-calibrating and the single-slope approaches. This can be explained by the spatial variations of palm biotissue optical properties that are much higher than the variations of a homogeneous tissue phantom.

Figure 5 provides the comparison of the average deviations in the extinction coefficient $\Delta\mu_{eff}^{P}$ for different types of perturbations in the studies of a biotissue phantom and a human palm. In Figure 5a the deviations in reconstructed extinction coefficient $\Delta\mu_{eff}^{P}$ for all types of perturbations in phantom studies are summarized. It can be seen from the diagram that deviations in the extinction coefficient calculated by the self-calibration approach are smaller than those calculated by the single-slope approach for all types of perturbations.

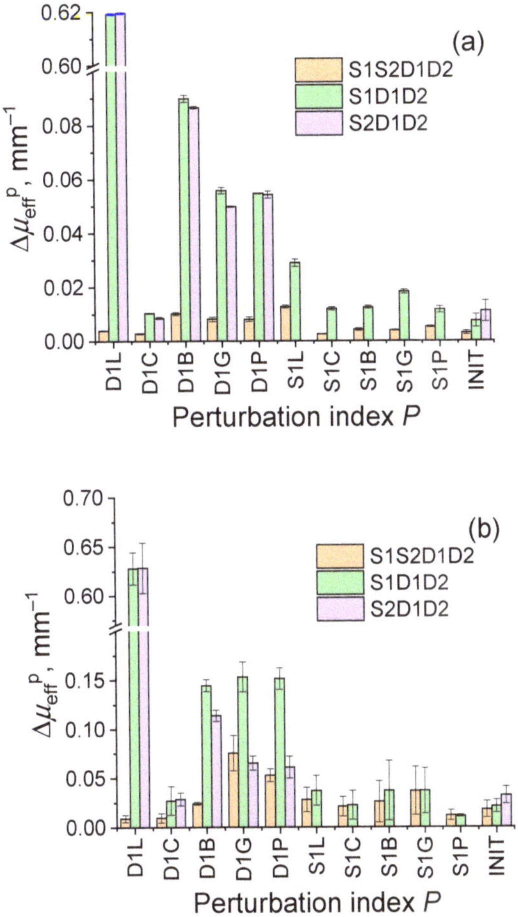


**Figure 5.** Deviations in extinction coefficient $\Delta\mu^{P}_{eff}$ spectra calculated for different types of perturbations by Equation (15) using single-slope approach in configurations S1D1D2 and S2D1D2 and self-calibrating approach S1S2D1D2 in phantom (**a**) and in vivo (**b**) studies. Error bars show deviations in $\Delta\mu^{P,m}_{eff}$ values.

For unperturbed measurements, deviations in the extinction coefficient are 0.003 mm$^{-1}$ for the self-calibration approach, and 0.007 and 0.011 mm$^{-1}$ for the single-slope technique in two configurations, respectively. For all types of perturbations, the deviation of the extinction coefficient calculated by the self-calibration approach does not exceed 0.013 mm$^{-1}$, while deviations in this value calculated using the single-slope approach exceed 0.6 mm$^{-1}$ for the loss perturbations in detection channels. It also can be seen from this plot that the deviations in the extinction coefficient caused by perturbations in the source channel are smaller than those applied to the detection channel. Figure 5b shows the values of the deviations for the in vivo measurements. Similar to Figure 5a, detector perturbations (D1L, D1C, D1B, D1G, D1P) lead to large values of deviations in the extinction coefficient calculated using the single-slope approach, and this effect is significantly lower for the self-calibrating approach. Larger error values in Figure 5b compared to Figure 5a can be explained by a smaller level of the DRS signals detected in vivo owing to higher extinction.

### 3.3. Reconstruction of Skin Chromophores and Scattering Properties

Examples of fitting the extinction spectra of the in vivo human palm calculated using the self-calibrating and single-slope approaches for unperturbed DRS data with expressions given by Equations (1), (2), and (10) using Equations (11)–(13) are shown in Figure 6. The fitting curve tracks the most pronounced visible features of the extinction spectrum: oxyhemoglobin peaks in the visible spectral range at 540 and 576 nm, deoxyhemoglobin peak at 756 nm, water absorption peak at 975 nm, and an overall decrease in the extinction coefficient from short to long wavelengths due to a decrease in absorption and scattering. However, there is some discrepancy between the reconstructed and experimentally obtained extinction spectra caused by the significant simplification of the applied model of light transfer in a human palm.

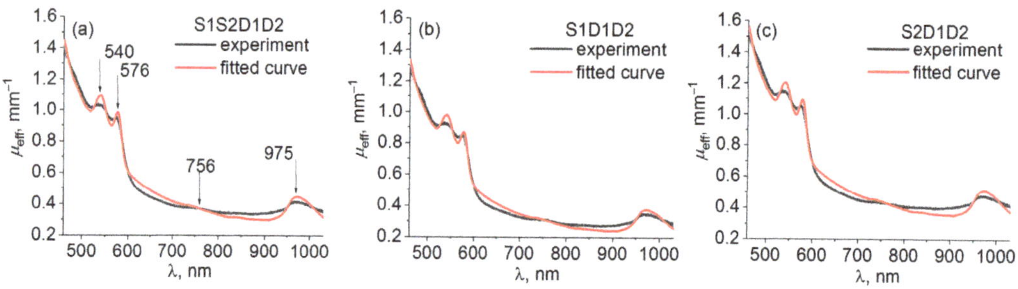


**Figure 6.** Fitting the extinction spectra of the in vivo human palm obtained from unperturbed measurements using (**a**) S1S2D1D2, (**b**) S1D1D2, and (**c**) S2D1D2 approaches with the reconstructed extinction spectra using Equations (11)–(13).

Figure 7 shows the values of blood $C_{blood}$ and water $C_{water}$ content, tissue oxygenation $StO_2$, and scattering properties $a$, $b$, and $f$ of human palm reconstructed from fitting the experimental extinction spectra which were obtained using different measurement approaches (self-calibrating S1S2D1D2 and single-slope S1D1D2, S2D1D2) from DRS data measured under different instrumental perturbations listed in Table 1. The reconstructed values obtained for unperturbed (INIT) data are in agreement with typical skin physiological parameters [18,43]: $StO_2$ is about 0.8 and $C_{blood}$ and $C_{water}$ are around 0.002 and 0.4, respectively. The reconstructed values of $a$, $b$, and $f$ yield the $\mu_s'(\lambda)$ dependence, which is in agreement with the reduced scattering spectra reported in [43]: the short wave range of the reconstructed spectrum tends to the typical values reported for epidermis owing to a smaller probing depth in this range, while in the NIR range the recovered spectrum corresponds well to the $\mu_s'$ spectrum reported for dermis.

As follows from the analysis of extinction coefficient deviations, skin optical parameters reconstructed from S1S2D1D2 extinction spectra demonstrate high stability for all DRS data obtained under all possible perturbations in which the deviation does not exceed 16%. In contrast, the parameters reconstructed from the single-slope data S1D1D2 and S2D1D2 demonstrate stability only for source perturbations and unperturbed data. These results are summarized in Figure 8, showing relative deviations in different skin characteristics obtained with all source (S1L, S1C, S1B, S1G, S1P) and all detector (D1L, D1C, D1B, D1G, D1P) perturbations. These plots demonstrate low average variations (less than 16%) of the parameters reconstructed from self-calibrating data for all types of perturbations, while detector perturbations may result in high variations (up to several times for particular perturbations) of reconstructed values from single-slope data (Figure 8a).

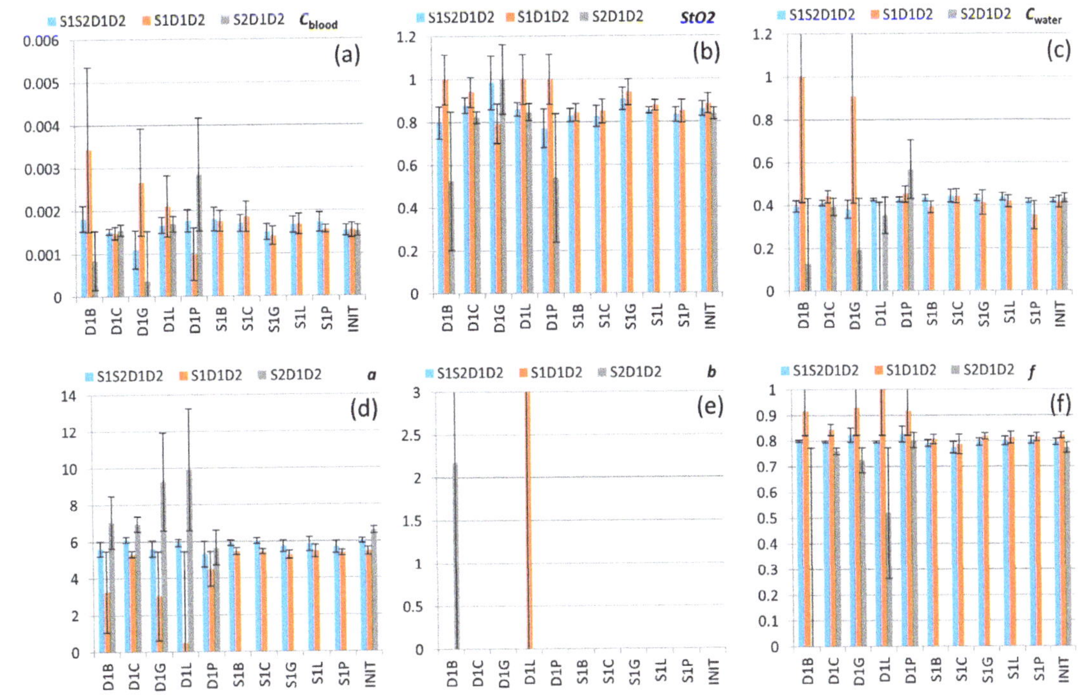


**Figure 7.** Values of $C_{blood}$ (**a**), $StO_2$ (**b**), $C_{water}$ (**c**), $a$ in mm$^{-1}$ (**d**), $b$ (**e**), and $f$ (**f**) reconstructed from the experimental extinction spectra using S1S2D1D2, S1D1D2, and S2D1D2 approaches and averaged over 3 measurements for each of the 10 perturbations (Table 1) and over 6 measurements for INIT data. Error bars show deviations in the corresponding reconstructed values in the series of the experiment. All values of $b$ are below $10^{-8}$, except the two corresponding to the perturbations D1B and D1L, which lead to largest deviations in the reconstructed extinction spectra from the unperturbed one (see Figure 4c,e).

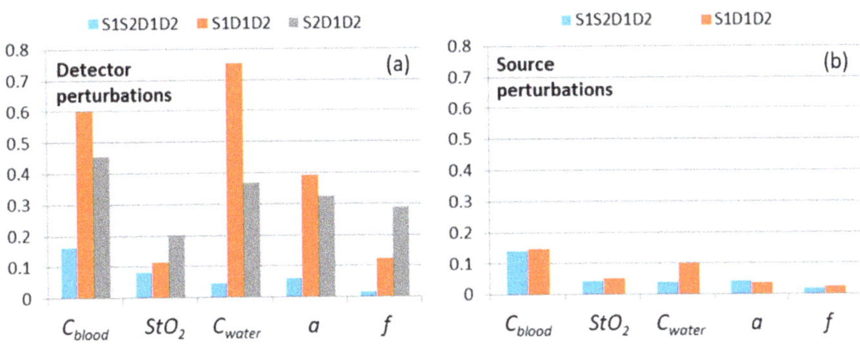


**Figure 8.** Relative deviations in $C_{blood}$, $StO_2$, $C_{water}$ , $a$, and $f$ values reconstructed using self-calibrating (S1S2D1D2) and single-slope (S1D1D2 and S2D1D2) approaches from the unperturbed measurement values. Relative deviations are averaged over different types of detector (**a**) and source (**b**) perturbations for each reconstructed value. Note that the plot (**b**) does not contain S2D1D2 data since perturbations introduced to the S1 channel provide no impact on these measurements.

## 4. Discussion

In this study we compared the capabilities of single- and dual-slope approaches in DRS to resist different perturbations that may occur during measurements. An experimental setup for wide-band DRS with a fiber-optic contact probe capable of employing a self-calibrating approach was constructed. This system contains a broadband fiber-optic source allowing for diffuse reflectance measurements in a wide VIS-NIR band (460–1030 nm). The upper wavelength range boundary is limited by the detector sensitivity curve, while detection in the short wavelength range is limited by strong probing light attenuation in biotissue. The self-calibrating scheme is based on symmetrical source-detector measurements performed through two fiber-optic switches for two source and two detection fibers of the probe (Figure 1). In order to fit the spectrometer dynamic range for the whole wavelength range for short (2 mm) and long (4 mm) SDDs, we applied different exposure times of 15 and 80 ms, respectively.

Different instrumental perturbations have been introduced into source and detector channels including attenuation, fiber bending, and corrupting probe–tissue interface in order to compare resistance to them of self-calibrating and single-slope approaches in phantom and in vivo studies. Both approaches have been applied to analyze the corresponding extinction spectrum deviations originating from the applied perturbation during DRS measurements. The results of phantom and in vivo studies have shown (Figures 3 and 4) that both approaches have resistance to instrumental perturbations introduced into the source channel (S1L, S1C, S1B, S1G, S1P). At the same time, perturbations introduced into the detection channel (D1L, D1C, D1B, D1G, D1P) may lead to significant deviations in the extinction spectra calculated by the single-slope approach (S1D1D2 or S2D1D2), while the self-calibrating approach (S1S2D1D2) demonstrated much higher resistance. This can be explained by the fact that Equation (5) for single slopes contains the ratio of detector transient characteristics and excludes source transient characteristics. However, Equations (6) and (7) are written under the assumption that the transient characteristics of both detectors are equal, therefore, perturbations introduced into one of the detector channels lead to the corruption of the evaluated extinction spectra. In contrast to this, in Equation (9) for the extinction spectrum calculated using self-calibrating approach, the transient functions for both sources and detectors are reduced. Figure 5a also demonstrates higher variations of extinction spectra calculated by the single-slope approach in comparison with the self-calibrating approach even for unperturbed data and all source perturbations. This effect is explained by residual instrumental perturbations (residual fiber bending, variations in SMA-connectors, etc.) that remained after perturbations introduced into source and detector channels during phantom studies. Figures 3f and 4f also demonstrate the imperfections in the detection channels of the designed DRS system seen by the discrepancies in the opposite sign between the values of $\mu_{eff}$ reconstructed from the unperturbed spectra in self-calibrating mode versus those for single-slope configurations S1D1D2 and S2D1D2.

Reconstruction of the biotissue properties from the obtained extinction coefficient spectra demonstrated that the self-calibrating approach provides reliable values with average deviations not exceeding 16% for all the considered perturbations. In this connection, if the spectral changes induced by instrumental perturbations in clinical conditions are similar to those we employed in this study, one can expect approximately the same accuracy of the self-calibrating technique. However, it should be noted that the DRS measurements in subcutaneous tissues are sensitive to the applied pressure of a DRS probe. Excessive pressure of a DRS probe may result in a significant change in tissue optical properties, while a loose probe–tissue contact may lead to a slight probe shift during the measurement procedure. Both effects may result in errors in the reconstructed physiological parameters. To avoid these effects, in the designed system, the DRS probe was equipped with a pressure control unit allowing it to keep the optimal pressure during measurements.

In the present study we applied the simplest reconstruction technique of skin optical properties based on a standard MATLAB minimization function, and the model of DRS spectra was taken from the diffusion approximation of radiation transfer theory for infinite and homogeneously scattering and absorbing medium. This model was applied to assess the instability of reconstructed tissue property values caused by instrumental variations; however, the reconstructed $\mu_{eff}$ spectrum has noticeable discrepancy with the experimental extinction spectrum obtained from the unperturbed measurements (Figure 6). We also limited the number of chromophores that contribute to the absorption spectrum in the described wavelength range and did not consider lipids because we focused on characterizing the chromophores of human palm, where the content of lipids is typically moderate and located mainly in hypodermis. Since our DRS system has SDDs of units in millimeters, it is mostly sensitive to superficial chromophores of skin such as water, blood, and melanin and to a lesser extent, to lipids in hypodermis at depths exceeding 1–2 mm. For a more precise reconstruction, the presence of the biotissue boundary and the skin layered structure should be taken into account, and more sophisticated algorithms of reconstruction should be used. It is essential for broadband DRS measurements in which light in the VIS and NIR spectral ranges penetrates to different depths in tissue. Monte-Carlo modeling of light transport can be used to take into account layered skin structure [44,45], and tissue properties can be derived using machine learning based on advanced theoretical and numerical models of light transport [46–49].

## 5. Conclusions

A comparative analysis of the sensitivity of single- and dual-slope (self-calibrating) approaches in DRS was performed using a custom-built wideband 460–1030 nm DRS setup in phantom and in vivo studies. Different instrumental perturbations have been introduced into source and detector channels in order to compare the stability of self-calibrating and single-slope approaches toward uncontrolled attenuations in individual channels, optical fiber bending, and optical inhomogeneities at the probe–tissue interface. Both single-slope and self-calibrating approaches have demonstrated high stability to perturbations introduced into the source channels. Perturbation in the detection channels may lead to significant deviations in the extinction spectra recovered from the measured back-reflectance spectra by the single-slope approach, however, the self-calibrating approach has demonstrated high stability for all types of perturbations. Reconstruction of the biotissue properties from the obtained extinction coefficient spectra demonstrated that the self-calibrating approach provides reliable values with average deviations not exceeding 16% for all the considered perturbations. Thus, we can conclude that the self-calibrating approach can be applied to DRS to provide robust measurements insensitive to instrumental perturbations in a wide VIS-NIR spectral band.

**Author Contributions:** Conceptualization, I.T., M.K. and E.S.; methodology, I.T., D.K. and E.S; software, O.S., V.P. and A.K. (Aleksandr Khilov); validation, E.S., M.K. and D.K.; formal analysis, E.S., M.K., I.T.; investigation, A.K. (Alexey Kostyuk), A.K. (Aleksandr Khilov), D.K., A.O.; resources, A.K. (Alexey Kostyuk) and I.T.; data curation, I.T., V.P, M.K. and E.S.; writing—original draft preparation, I.T. and V.P.; writing—review and editing, I.T., V.P., M.K. and E.S.; visualization, I.T., V.P. and A.K. (Aleksandr Khilov). project administration, I.T.; All authors have read and agreed to the published version of the manuscript.

**Funding:** The authors acknowledge the support by Center of Excellence «Center of Photonics» funded by The Ministry of Science and Higher Education of the Russian Federation, Contract No. 075-15-2022-316.

**Institutional Review Board Statement:** Not applicable.

**Informed Consent Statement:** Informed consent was obtained from all subjects involved in the study.

**Data Availability Statement:** The data presented in this study are available on a reasonable request from the corresponding author.

**Acknowledgments:** The authors thank Vladimir Vorobjev and Sergey Pozhidaev for the electrical and mechanical engineering of the developed DRS setup; we also thank Mikhail Kleshnin for valuable recommendations during discussion about probe design.

**Conflicts of Interest:** The authors declare no conflict of interest.

## References

1. Pinti, P.; Tachtsidis, I.; Hamilton, A.; Hirsch, J.; Aichelburg, C.; Gilbert, S.; Burgess, P.W. The present and future use of functional near-infrared spectroscopy (fnirs) for cognitive neuroscience. *Ann. N. Y. Acad. Sci.* **2020**, *1464*, 5–29. [CrossRef]
2. De Boer, L.; Molenkamp, B.; Bydlon, T.; Hendriks, B.; Wesseling, J.; Sterenborg, H.; Ruers, T. Fat/water ratios measured with diffuse reflectance spectroscopy to detect breast tumor boundaries. *Breast Cancer Res. Treat.* **2015**, *152*, 509–518. [CrossRef] [PubMed]
3. Soares, J.S.; Barman, I.; Dingari, N.C.; Volynskaya, Z.; Liu, W.; Klein, N.; Plecha, D.; Dasari, R.R.; Fitzmaurice, M. Diagnostic power of diffuse reflectance spectroscopy for targeted detection of breast lesions with microcalcifications. *Proc. Natl. Acad. Sci. USA* **2013**, *110*, 471–476. [CrossRef] [PubMed]
4. Tromberg, B.J.; Zhang, Z.; Leproux, A.; O'Sullivan, T.D.; Cerussi, A.E.; Carpenter, P.M.; Mehta, R.S.; Roblyer, D.; Yang, W.; Paulsen, K.D. Predicting responses to neoadjuvant chemotherapy in breast cancer: Acrin 6691 trial of diffuse optical spectroscopic imagingoptical imaging of breast cancer chemotherapy response. *Cancer Res.* **2016**, *76*, 5933–5944. [CrossRef]
5. Zonios, G.; Dimou, A.; Bassukas, I.; Galaris, D.; Tsolakidis, A.; Kaxiras, E. Melanin absorption spectroscopy: New method for noninvasive skin investigation and melanoma detection. *J. Biomed. Opt.* **2008**, *13*, 014017. [CrossRef]
6. Bolton, F.J.; Bernat, A.S.; Bar-Am, K.; Levitz, D.; Jacques, S. Portable, low-cost multispectral imaging system: Design, development, validation, and utilization. *J. Biomed. Opt.* **2018**, *23*, 121612. [CrossRef]
7. Potapova, E.; Dremin, V.; Zherebtsov, E.; Makovik, I.; Zherebtsova, A.; Dunaev, A.; Podmasteryev, K.; Sidorov, V.; Krupatkin, A.; Khakhicheva, L. Evaluation of microcirculatory disturbances in patients with rheumatic diseases by the method of diffuse reflectance spectroscopy. *Hum. Physiol.* **2017**, *43*, 222–228. [CrossRef]
8. Hsu, C.-K.; Tzeng, S.-Y.; Yang, C.-C.; Lee, J.Y.-Y.; Huang, L.L.-H.; Chen, W.-R.; Hughes, M.; Chen, Y.-W.; Liao, Y.-K.; Tseng, S.-H. Non-invasive evaluation of therapeutic response in keloid scar using diffuse reflectance spectroscopy. *Biomed. Opt. Express* **2015**, *6*, 390–404. [CrossRef]
9. de Koning, S.G.B.; Baltussen, E.J.; Karakullukcu, M.B.; Dashtbozorg, B.; Smit, L.A.; Dirven, R.; Hendriks, B.H.; Sterenborg, H.J.; Ruers, T.J. Toward complete oral cavity cancer resection using a handheld diffuse reflectance spectroscopy probe. *J. Biomed. Opt.* **2018**, *23*, 121611. [CrossRef]
10. Spliethoff, J.W.; Evers, D.J.; Klomp, H.M.; van Sandick, J.W.; Wouters, M.W.; Nachabe, R.; Lucassen, G.W.; Hendriks, B.H.; Wesseling, J.; Ruers, T.J. Improved identification of peripheral lung tumors by using diffuse reflectance and fluorescence spectroscopy. *Lung Cancer* **2013**, *80*, 165–171. [CrossRef]
11. Evers, D.; Nachabe, R.; Hompes, D.; Van Coevorden, F.; Lucassen, G.; Hendriks, B.; van Velthuysen, M.-L.; Wesseling, J.; Ruers, T. Optical sensing for tumor detection in the liver. *Eur. J. Surg. Oncol. (EJSO)* **2013**, *39*, 68–75. [CrossRef] [PubMed]
12. Tanis, E.; Evers, D.J.; Spliethoff, J.W.; Pully, V.V.; Kuhlmann, K.; van Coevorden, F.; Hendriks, B.H.; Sanders, J.; Prevoo, W.; Ruers, T.J. In vivo tumor identification of colorectal liver metastases with diffuse reflectance and fluorescence spectroscopy. *Lasers Surg. Med.* **2016**, *48*, 820–827. [CrossRef] [PubMed]
13. Voulgarelis, S.; Fathi, F.; Stucke, A.G.; Daley, K.D.; Kim, J.; Zimmerman, M.A.; Hong, J.C.; Starkey, N.; Allen, K.P.; Yu, B. Evaluation of visible diffuse reflectance spectroscopy in liver tissue: Validation of tissue saturations using extracorporeal circulation. *J. Biomed. Opt.* **2021**, *26*, 055002. [CrossRef] [PubMed]
14. Nogueira, M.S.; Maryam, S.; Amissah, M.; Lu, H.; Lynch, N.; Killeen, S.; O'Riordain, M.; Andersson-Engels, S. Evaluation of wavelength ranges and tissue depth probed by diffuse reflectance spectroscopy for colorectal cancer detection. *Sci. Rep.* **2021**, *11*, 798. [CrossRef] [PubMed]
15. Baltussen, E.J.; Brouwer De Koning, S.G.; Hendriks, B.H.; Jóźwiak, K.; Sterenborg, H.J.; Ruers, T.J. Comparing in vivo and ex vivo fiberoptic diffuse reflectance spectroscopy in colorectal cancer. *Transl. Biophotonics* **2019**, *1*, e201900008. [CrossRef]
16. Konugolu Venkata Sekar, S.; Farina, A.; Dalla Mora, A.; Lindner, C.; Pagliazzi, M.; Mora, M.; Aranda, G.; Dehghani, H.; Durduran, T.; Taroni, P. Broadband (550–1350 nm) diffuse optical characterization of thyroid chromophores. *Sci. Rep.* **2018**, *8*, 10015. [CrossRef]
17. Lanka, P.; Segala, A.; Farina, A.; Sekar, S.K.V.; Nisoli, E.; Valerio, A.; Taroni, P.; Cubeddu, R.; Pifferi, A. Non-invasive investigation of adipose tissue by time domain diffuse optical spectroscopy. *Biomed. Opt. Express* **2020**, *11*, 2779–2793. [CrossRef]
18. Jacques, S.L. Optical properties of biological tissues: A review. *Phys. Med. Biol.* **2013**, *58*, R37. [CrossRef]
19. Hou, J.; Ness, S.S.; Tschudi, J.; O'Farrell, M.; Veddegjerde, R.; Martinsen, Ø.G.; Tønnessen, T.I.; Strand-Amundsen, R. Assessment of intestinal ischemia–reperfusion injury using diffuse reflectance vis-nir spectroscopy and histology. *Sensors* **2022**, *22*, 9111. [CrossRef]
20. Kraitl, J.; Timm, U.; Ewald, H. Non-Invasive Measurement of Blood and Tissue Parameters Based on Vis-Nir Spectroscopy. In Proceedings of the Optical Diagnostics and Sensing XIII: Toward Point-of-Care Diagnostics; Volume 8591, pp. 33–40. [CrossRef]

21. Turchin, I.V. Methods of biomedical optical imaging: From subcellular structures to tissues and organs. *Phys. Uspekhi* **2016**, *59*, 487. [CrossRef]
22. Tseng, S.-H.; Bargo, P.; Durkin, A.; Kollias, N. Chromophore concentrations, absorption and scattering properties of human skin in-vivo. *Opt. Express* **2009**, *17*, 14599–14617. [CrossRef] [PubMed]
23. Scholkmann, F.; Metz, A.J.; Wolf, M. Measuring tissue hemodynamics and oxygenation by continuous-wave functional near-infrared spectroscopy—How robust are the different calculation methods against movement artifacts? *Physiol. Meas.* **2014**, *35*, 717. [CrossRef] [PubMed]
24. Merritt, S.; Gulsen, G.; Chiou, G.; Chu, Y.; Deng, C.; Cerussi, A.E.; Durkin, A.J.; Tromberg, B.J.; Nalcioglu, O. Comparison of water and lipid content measurements using diffuse optical spectroscopy and mri in emulsion phantoms. *Technol. Cancer Res. Treat.* **2003**, *2*, 563–569. [CrossRef] [PubMed]
25. Yu, B.; Fu, H.; Bydlon, T.; Bender, J.E.; Ramanujam, N. Diffuse reflectance spectroscopy with a self-calibrating fiber optic probe. *Opt. Lett.* **2008**, *33*, 1783–1785. [CrossRef]
26. Kleshnin, M.S.; Turchin, I.y.V. Evaluation of oxygenation in the surface layers of biological tissues based on diffuse optical spectroscopy with automated calibration of measurements. *Quantum Electron.* **2019**, *49*, 628. [CrossRef]
27. Hueber, D.M.; Fantini, S.; Cerussi, A.E.; Barbieri, B.B. New optical probe designs for absolute (self-calibrating) nir tissue hemoglobin measurements. In Proceedings of the Optical Tomography and Spectroscopy of Tissue III; Volume 3597, pp. 618–631. [CrossRef]
28. Feng, S.; Zeng, F.-A.; Chance, B. Photon migration in the presence of a single defect: A perturbation analysis. *Appl. Opt.* **1995**, *34*, 3826–3837. [CrossRef]
29. Sassaroli, A.; Blaney, G.; Fantini, S. Dual-slope method for enhanced depth sensitivity in diffuse optical spectroscopy. *JOSA A* **2019**, *36*, 1743–1761. [CrossRef]
30. Fantini, S.; Blaney, G.; Sassaroli, A. Transformational change in the field of diffuse optics: From going bananas to going nuts. *J. Innov. Opt. Health Sci.* **2020**, *13*, 1930013. [CrossRef]
31. Kleiser, S.; Ostojic, D.; Nasseri, N.; Isler, H.; Bucher, H.U.; Bassler, D.; Wolf, M.; Scholkmann, F.; Karen, T. In vivo precision assessment of a near-infrared spectroscopy-based tissue oximeter (oxyprem v1. 3) in neonates considering systemic hemodynamic fluctuations. *J. Biomed. Opt.* **2018**, *23*, 067003. [CrossRef]
32. Xu, R.X.; Qiang, B.; Mao, J.J.; Povoski, S.P. Development of a handheld near-infrared imager for dynamic characterization of in vivo biological tissue systems. *Appl. Opt.* **2007**, *46*, 7442–7451. [CrossRef]
33. Wang, B.; Povoski, S.P.; Cao, X.; Sun, D.; Xu, R.X. Dynamic schema for near infrared detection of pressure-induced changes in solid tumors. *Appl. Opt.* **2008**, *47*, 3053–3063. [CrossRef] [PubMed]
34. Mesradi, M.; Genoux, A.; Cuplov, V.; Abi-Haidar, D.; Jan, S.; Buvat, I.; Pain, F. Experimental and analytical comparative study of optical coefficient of fresh and frozen rat tissues. *J. Biomed. Opt.* **2013**, *18*, 117010. [CrossRef] [PubMed]
35. Bevilacqua, F.; Berger, A.J.; Cerussi, A.E.; Jakubowski, D.; Tromberg, B.J. Broadband absorption spectroscopy in turbid media by combined frequency-domain and steady-state methods. *Appl. Opt.* **2000**, *39*, 6498–6507. [CrossRef] [PubMed]
36. Blaney, G.; Donaldson, R.; Mushtak, S.; Nguyen, H.; Vignale, L.; Fernandez, C.; Pham, T.; Sassaroli, A.; Fantini, S. Dual-slope diffuse reflectance instrument for calibration-free broadband spectroscopy. *Appl. Sci.* **2021**, *11*, 1757. [CrossRef] [PubMed]
37. Pham, T.; Fernandez, C.; Blaney, G.; Tgavalekos, K.; Sassaroli, A.; Cai, X.; Bibu, S.; Kornbluth, J.; Fantini, S. Noninvasive optical measurements of dynamic cerebral autoregulation by inducing oscillatory cerebral hemodynamics. *Front. Neurol.* **2021**, *12*, 745987. [CrossRef] [PubMed]
38. Zonios, G.; Dimou, A. Modeling diffuse reflectance from semi-infinite turbid media: Application to the study of skin optical properties. *Opt. Express* **2006**, *14*, 8661–8674. [CrossRef]
39. Ishimaru, A. *Wave Propagation and Scattering in Random Media*; Academic Press: New York, NY, USA, 1978; Volume 2. [CrossRef]
40. Patterson, M.S.; Schwartz, E.; Wilson, B.C. Quantitative reflectance spectrophotometry for the noninvasive measurement of photosensitizer concentration in tissue during photodynamic therapy. In Proceedings of the Photodynamic Therapy: Mechanisms; Volume 1065, pp. 115–122. [CrossRef]
41. Nachabé, R.; Evers, D.J.; Hendriks, B.H.; Lucassen, G.W.; van der Voort, M.; Wesseling, J.; Ruers, T.J. Effect of bile absorption coefficients on the estimation of liver tissue optical properties and related implications in discriminating healthy and tumorous samples. *Biomed. Opt. Express* **2011**, *2*, 600–614. [CrossRef]
42. Beard, P. Biomedical photoacoustic imaging. *Interface Focus* **2011**, *1*, 602–631. [CrossRef]
43. Shimojo, Y.; Nishimura, T.; Hazama, H.; Ozawa, T.; Awazu, K. Measurement of absorption and reduced scattering coefficients in asian human epidermis, dermis, and subcutaneous fat tissues in the 400-to 1100-nm wavelength range for optical penetration depth and energy deposition analysis. *J. Biomed. Opt.* **2020**, *25*, 045002. [CrossRef]
44. Kurakina, D.; Perekatova, V.; Sergeeva, E.; Kostyuk, A.; Turchin, I.; Kirillin, M. Probing depth in diffuse reflectance spectroscopy of biotissues: A monte carlo study. *Laser Phys. Lett.* **2022**, *19*, 035602. [CrossRef]
45. Zherebtsov, E.; Dremin, V.; Popov, A.; Doronin, A.; Kurakina, D.; Kirillin, M.; Meglinski, I.; Bykov, A. Hyperspectral imaging of human skin aided by artificial neural networks. *Biomed. Opt. Express* **2019**, *10*, 3545–3559. [CrossRef] [PubMed]
46. Kleshnin, M.S. Deep learning neural network estimation of tissue oxygenation based on diffuse optical spectroscopy. *Laser Phys.* **2019**, *29*, 085603. [CrossRef]

47. Fernandes, L.; Carvalho, S.; Carneiro, I.; Henrique, R.; Tuchin, V.V.; Oliveira, H.P.; Oliveira, L.M. Diffuse reflectance and machine learning techniques to differentiate colorectal cancer ex vivo. *Chaos Interdiscip. J. Nonlinear Sci.* **2021**, *31*, 053118. [CrossRef] [PubMed]
48. Fredriksson, I.; Larsson, M.; Strömberg, T. Machine learning for direct oxygen saturation and hemoglobin concentration assessment using diffuse reflectance spectroscopy. *J. Biomed. Opt.* **2020**, *25*, 112905. [CrossRef]
49. Nguyen, M.H.; Zhang, Y.; Wang, F.; Linan, J.D.L.G.E.; Markey, M.K.; Tunnell, J.W. Machine learning to extract physiological parameters from multispectral diffuse reflectance spectroscopy. *J. Biomed. Opt.* **2021**, *26*, 052912. [CrossRef]

**Disclaimer/Publisher's Note:** The statements, opinions and data contained in all publications are solely those of the individual author(s) and contributor(s) and not of MDPI and/or the editor(s). MDPI and/or the editor(s) disclaim responsibility for any injury to people or property resulting from any ideas, methods, instructions or products referred to in the content.

diagnostics

MDPI

Article

# The Glycerol-Induced Perfusion-Kinetics of the Cat Ovaries in the Follicular and Luteal Phases of the Cycle

**Alexey A. Selifonov [1,*], Andrey S. Rykhlov [2] and Valery V. Tuchin [3]**

1 Education and Research Institute of Nanostructures and Biosystems, Saratov State University, Saratov 410012, Russia
2 Clinic "Veterinary Hospital", Saratov State University of Genetics, Biotechnology and Engineering Named after N.I. Vavilov, Saratov 410012, Russia
3 Science Medical Center, Saratov State University, Saratov 410012, Russia
* Correspondence: peshka029@gmail.com; Tel.: +7-987-356-40-30

**Abstract:** The method of immersion optical clearing reduces light scattering in tissues, which improves the use of optical technologies in the practice of clinicians. In this work, we studied the optical and molecular diffusion properties of cat ovarian tissues in the follicular (F-ph) and luteal (L-ph) phases under the influence of glycerol using reflectance spectroscopy in a broad wavelength range from 200 to 800 nm. It was found that the reflectance and transmittance of the ovaries are significantly lower in the range from 200 to 600 nm than for longer wavelengths from 600 to 800 nm, and the efficiency of optical clearing is much lower for the ovaries in the luteal phase compared to the follicular phase. For shorter wavelengths, the following tissue transparency windows were observed: centered at 350 nm and wide (46 ± 5) nm, centered at 500 nm and wide (25 ± 7) nm for the F-ph state and with a center of 500 nm and a width of (21 ± 6) nm for the L-ph state. Using the free diffusion model, Fick's law of molecular diffusion and the Bouguer–Beer–Lambert radiation attenuation law, the glycerol/tissue water diffusion coefficient was estimated as $D = (1.9 \pm 0.2)10^{-6}$ cm$^2$/s for ovaries at F-ph state and $D = (2.4 \pm 0.2)10^{-6}$ cm$^2$/s—in L-ph state, and the time of complete dehydration of ovarian samples, 0.8 mm thick, as 22.3 min in F-ph state and 17.7 min in L-ph state. The ability to determine the phase in which the ovaries are stated, follicular or luteal, is also important in cryopreservation, new reproductive technologies and ovarian implantation.

**Keywords:** ovarian tissues; follicular phase; luteal phase; glycerol; tissue water; total transmittance spectra; diffuse reflectance spectra; diffusion coefficient; optical clearing efficiency

**Citation:** Selifonov, A.A.; Rykhlov, A.S.; Tuchin, V.V. The Glycerol-Induced Perfusion-Kinetics of the Cat Ovaries in the Follicular and Luteal Phases of the Cycle. *Diagnostics* **2023**, *13*, 490. https://doi.org/10.3390/diagnostics13030490

Academic Editor: Viktor Dremin

Received: 23 December 2022
Revised: 25 January 2023
Accepted: 26 January 2023
Published: 29 January 2023

**Copyright:** © 2023 by the authors. Licensee MDPI, Basel, Switzerland. This article is an open access article distributed under the terms and conditions of the Creative Commons Attribution (CC BY) license (https://creativecommons.org/licenses/by/4.0/).

## 1. Introduction

Every year, around the world, there is an increase in the number of diagnosed oncological diseases, including among patients of reproductive age [1]. For example, in the United States, about 70,000 cancer patients under the age of 45 are diagnosed annually [2]. Patients with malignant neoplasms, according to existing modern medical standards, undergo complex chemotherapy and radiation therapy. As a result of such treatment, there is a high probability of partial or complete loss of fertility in women, due to the high cytotoxicity of antitumor treatment [3–6]. A large group of patients is young women and girls, whose treatment requires bone marrow transplantation, before which alkylating drugs are used in high concentrations, which in most cases leads to sterilization. Currently, it is possible to preserve the reproductive function of women with cancer, with impaired reproductive function or with premature ovarian failure; this is the cryopreservation (freezing) of healthy ovarian tissue with subsequent transplantation or autotransplantation after recovery [7–10].

One of the most informative and reliable methods for examining ovarian tissue is laparoscopy, which is widely used in gynecology for both diagnostic and surgical purposes [11,12]. The laparoscope is inserted into the abdominal cavity through a small incision, which allows one to directly examine the organs of the small pelvis and abdominal cavity or, by connecting a video camera, transmit the image to the monitor. However, such an image is formed only by light reflection from the surface of the organ under study; the internal structure of the organ is hidden from observation due to strong light scattering by the tissues of this organ. Using the immersion optical clearing of tissue, it is possible to suppress scattering and observe previously hidden pathological changes in tissues for some time [13–17]. Typical times for the optical clearing of the upper layers of the tissue are 20–40 min. It is important that radiocontrast and MRI contrast agents can act as optical clearing agents. This opens the way to multimodal diagnostics and the support of laparoscopic surgery. It should be noted that hysterosalpingography and hysteroscopy [11,12,18] can be combined in one study with the possibility of obtaining high-quality optical images using laparoscopically compatible optical imaging techniques, as radiopaque and MRI agents are good optical clearing agents [13–17].

Cryopreservation is a new method that is successfully used in clinical practice. By 2020, about 85 transplantations of cryopreserved ovarian tissue have been performed, and cases of birth of 30 children have been described, although a longer follow-up of patients is required [19,20]. The standard method for cryopreservation of ovarian tissue is slow freezing using a medium with the addition of cryoprotectants: dimethyl sulfoxide (DMSO), ethylene glycol and 1,2-propanediol (PrOH), which are able to penetrate cell membranes and provide their protection. Slow freezing is carried out with liquid nitrogen for several hours [14,21,22]. However, with slow freezing, there is a risk of damage to cells by ice crystals [8,23,24]. Therefore, penetrating cryoprotectants are often used in combination with non-penetrating ones such as sucrose, glycerol, or human serum albumin [22]. They protect cells through dehydration and stabilization of the lipid layer and proteins. The use of glycerol during the cryopreservation of ovarian tissue can be useful for maintaining the viability of follicles, as glycerol causes the dehydration of cells and, when mixed with water, reduces the freezing point (the temperature of ice formation in cells and solutions) and increases the viscosity of aqueous solutions. After preparing the biological material by soaking it in cryogenic liquids, the vitrification procedure is used, which consists of immersing the material in liquid nitrogen. This method effectively preserves the morphology and viability of the follicles [25–27].

The method of the cryopreservation of ovarian tissue is new and requires further comprehensive study. Many researchers estimate the recovery of ovarian function after cryopreserved tissue transplantation as short-term. Ischemia that occurs in the first hours after transplantation can lead to the death of more than a third of primordial follicles and, therefore, is the main reason for the decrease in the functional activity of the ovary [28]. To restore the reproductive potential, it is extremely important to reduce the time interval of ischemia and accelerate the revascularization of grafts. In this regard, the transplantation of whole ovarian tissue on a vascular pedicle has been proposed as the most acceptable approach compared to the transplantation of the ovarian cortex alone [29,30]. As shown in animals, the transplantation of an intact whole ovary with microsurgical vascular anastomosis, despite the technical complexity of this procedure, is the only solution to the problem, as it provides a direct blood supply to the ovarian tissue after transplantation, minimizing the risk of ischemia. Therefore, the whole ovary cryopreservation method requires development and further research. The determination of the kinetic parameters of ovarian perfusion with cryoprotectants in different phases of the cycle, including the rate of diffusion of glycerol and the flow of tissue water induced by it, as well as changes in the optical properties of ovarian tissue, is a necessary condition for creating personalized clinical protocols for cryopreservation.

The aim of this work was to study the perfusion-kinetic properties of cat ovaries in the follicular (F-ph) and luteal (L-ph) phases of the cycle by optical clearing method under the influence of glycerol and using diffuse reflectance spectroscopy.

## 2. Materials and Methods

### 2.1. The Structure of the Ovaries and the Cycle

The ovaries are a paired organ located on the sides of the uterus, next to the ampullar sections of the fallopian tubes, their size in women ranges from 1.5 to 5 cm [31] and in cats, from 0.5 to 1.5 cm [32]. From above, the ovaries are covered with a layer of the epithelium; the next layer consists of connective tissue and contains many elastic fibers (Figure 1). The medulla contains many blood vessels and nerve endings. In the cortical layer, there are follicles in which eggs are formed and mature.

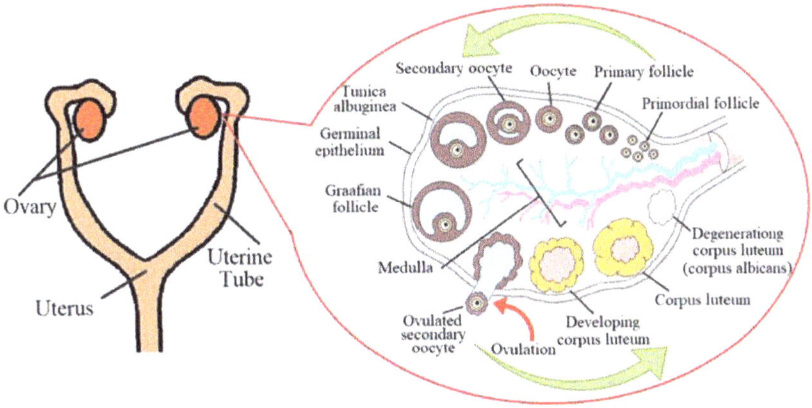


**Figure 1.** Schematic representation of the ovarian cycle. Adopted from Ref. [31].

The female cycle normally lasts from 21 to 35 days. The main phases can be distinguished: 1: follicular (F-ph), in this phase of the cycle, there are many growing and primary follicles in the ovaries (Figure 1); 2: ovulation is a hormone-dependent process of the rupture of the wall of the tertiary (preovulatory) follicle and the release of the female germ cell, and a peak level of hormones is observed: follitropin and lutropin; and 3: luteal phase (L-ph). The corpus luteum has a size of 1.0 to 2.7 cm. A gland of temporary secretion contains lutein and produces a large amount of progesterone, the dominant hormone of the luteal phase, which is important for the safety and proper development of a potential embryo [33].

When studying the blood supply to the ovaries of cows, it was found that the uterine branch of the ovarian artery and especially its anastomosis with the uterine artery were larger on the side of the ovary containing the corpus luteum [34]. The blood supply of the mature corpus luteum is the highest of all body organs per unit volume of tissue. An increase in blood supply is an integral part of the development of the corpus luteum. This important process, mediated by angiogenic growth factors, includes the destruction of the basement membrane of the follicles, the proliferation and migration of endothelial cells, and the development of a large network of capillaries [35]. The corpus luteum functions for only a few (4–7) days and then undergoes involution. A white body (scar) appears in its place [36] (Figure 1).

### 2.2. Histological Examination

Normal ovarian tissue was taken from outbred cats aged 1 to 12 years with a diagnosis of "clinically healthy". The ovaries were collected after laparoscopic oophorectomy and ovariohysterectomy from 10 cats. Animals were administered general anesthesia: premedication—meditin (0.1%); intravenous anesthesia—zoletil 100; and alpha 2—antagonist antiemetic for withdrawal from anesthesia. According to visual inspection, all ovaries planned for the study were divided into two groups: "light" and "dark"

(Figure 2). Ten light and ten dark tissue samples from different individuals were used for histological examination.

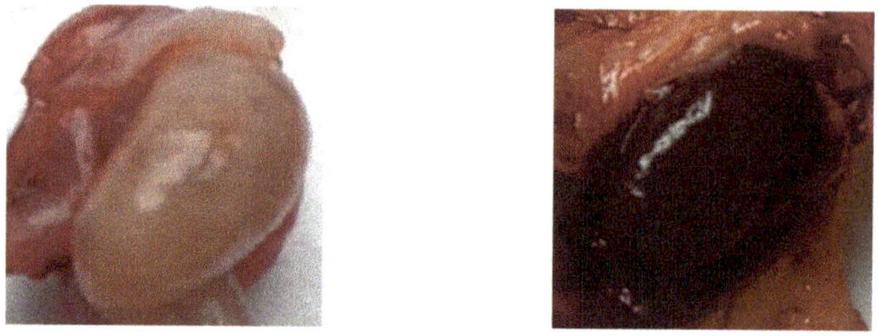

(a) (b)

**Figure 2.** Photos of the studied clinically healthy cat ovaries: (**a**) light; (**b**) dark.

In vitro histological studies were carried out using halves of each ovary, which were manually cut with a scalpel and fixed. The other halves of the ovaries without fixation were kept frozen for ex vivo optical measurements. The material for histological examination was prepared after no more than 48 h had passed after oophorectomy and ovariohysterectomy; 10% buffered formalin was used for tissue fixation. The thickness of tissue sections was 2–3 μm. The hematoxylin–eosin staining method was used for the histological examination of all samples.

To obtain histological scans, an Aperio AT2 digital slide converter (on-screen diagnostic scanner) equipped with an LED light source and calibration tools was used. According to the results of the histological studies, it was proved that normal, pathologically unchanged tissues were selected for research, that dark samples contain the corpus luteum and correspond to the luteal phase, and light samples contain multiple follicles, which corresponds to the follicular phase.

### 2.3. *Optical Measurements*

Measurements of the optical properties of cat ovarian tissue were performed ex vivo without tissue fixation. The thickness of sections (samples) of tissue was measured with an electronic micrometer (Union Source CO., Ltd., Ningbo, China). The measurements were carried out at several points of the sample and averaged. The accuracy of each measurement was ±0.1 mm. The thickness of the tissue section of both dark and light ovaries averaged (0.8 ± 0.1) mm. To measure the diffuse reflectance spectra (DRS) and the total transmittance spectra (TTS) of the tissue samples in the spectral range of 200–800 nm, a Shimadzu UV-2550 double-beam spectrophotometer (Tokyo, Japan) with an integrating sphere was used (Figure 2). A total of 20 ovaries, 10 light and 10 dark, were examined for optical measurements. To study the kinetics of the DRS, ten sections were taken from every five light and five dark samples. Similarly, ten sections from the other five + five ovaries were used to measure the TTS.

The radiation source was a halogen lamp with radiation filtering in the studied spectral range. The limiting resolution of the spectrometer was 0.1 nm. Prior to measurements, the spectra were normalized using a $BaSO_4$ reference reflector with a suitable reflectivity for the entire spectral range, including UV. All measurements were carried out at room temperature (~25 °C) and normal atmospheric pressure. Each sample of the studied tissue was fixed with a double-sided adhesive tape in a special frame with a window of 0.5 × 0.5 cm in a quartz cuvette so that the tissue sample was pressed against the wall of the cuvette and subjected to optical measurement of DRS or TTS as shown in Figure 3. To measure the TTS, a quartz cuvette with a tissue sample was placed directly in front of the integrating sphere, collecting all the radiation transmitted through the tissue sample. The diameter of the light beam incident on the sample was 3 mm. The initial DRS or TTS spectrum was taken from the ovarian tissue sample pressed against the cuvette wall. Then, glycerol was added into the space between the sample surface and the cuvette wall, after which, measurements were carried out for 100 min until the time dependence was saturated due to the completion of the glycerol/interstitial water diffusion process. The measurement of the DRS kinetics was used for the determination of the diffusion coefficient of the molecular flux induced by the topical application of glycerol to a tissue sample. In the study, a chemically pure 99.5%-glycerol was used (Akrihimfarm LLC., Moscow, Russia).

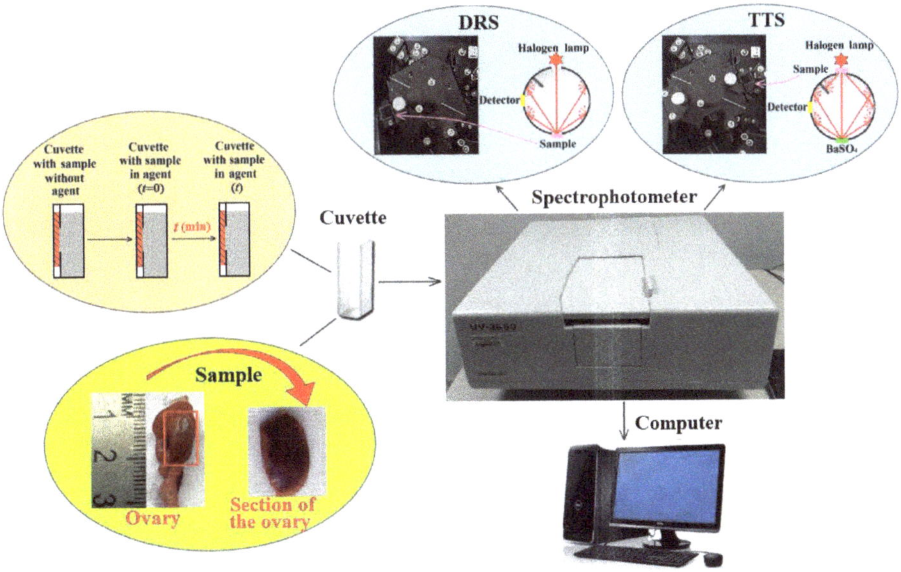


**Figure 3.** Scheme of the experimental setup for measuring the DRS and TTS of ex vivo samples of cat ovarian tissue.

It is important to note that, in the wavelength range from 150 to 800 nm, the absorption of glycerol is negligible [37]. However, to determine the effectiveness of optical clearing after the clearing process was completed, the frame with the sample was transferred to a similar, but dry, clean cuvette. Then, the final values of the DRS and TTS of the sample were measured and compared with the initial values before clearing, which were also obtained in a cuvette without glycerol.

## 3. Calculations

The determination of the diffusion coefficient of glycerol/interstitial water in tissue is based on measuring the kinetics of the DRS. Figure 4 illustrates this schematically. The process of glycerol/interstitial water transport in a sample can be described in terms of the model of free diffusion [13–17,38–40]. Geometrically, a sample of tissue can be represented as a plane-parallel plate of finite thickness. Using the second Fick's law and performing transformations based on the use of the modified Bouguer–Beer–Lambert law, described in detail in refs. [16,38], we obtain an expression for the difference $\Delta A(t, \lambda)$ between the effective optical density at the current time $A(t, \lambda)$ and at the initial time $A(t = 0, \lambda)$:

$$\Delta A(t, \lambda) = A(t, \lambda) - A(t = 0, \lambda) = \Delta\mu_{eff}(t, \lambda)L \sim C_0\{1 - \exp(-t/\tau)\}L \quad (1)$$

$$I = I_0 \exp[-\mu_{eff}L], \ \mu_{eff}(t, \lambda) = \sqrt{3\mu_a(\mu_a + \mu_s')} \rightarrow \Delta\mu_{eff}(t, \lambda)$$

where the effective optical density is determined from the measurements of DRS:

$$A = -\log R(t, \lambda), \quad (2)$$

$$\tau = \frac{4l^2}{\pi^2 D}, \tag{3}$$

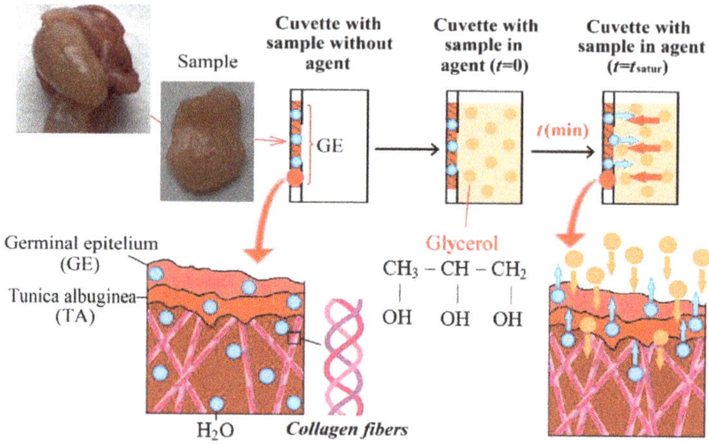


**Figure 4.** Diagram showing the interaction of a hyperosmotic agent (glycerol) with ovarian tissue.

$t$ is the time in seconds during which the diffusion occurs, $\lambda$ is the wavelength in nm, $\Delta\mu_{eff}$ $(t, \lambda)$ is the difference between the effective coefficient of attenuation of light in the tissue $\mu_{eff}(t, \lambda)$ at the current time and at the initial time, 1/cm; $L$ is the average path length of photons, which in the backscattering mode is $L \cong 2l_d$, $(l_d)^{-1} = \mu_{eff}$; $\mu'_s = \mu_s(1 - g)$, 1/cm; $g$ is the scattering anisotropy factor (varies from 0 to 1, for many tissues, $g \cong 0.93$) [16,39]; and for transmission mode $L \cong l$, $l$ is the thickness of the sample, cm; $D$ is the diffusion coefficient of the glycerol/interstitial water molecules, $cm^2/s$; and $C_0$ is the initial concentration of the glycerol, mol/L.

The recorded DRSs [$R(\lambda)$, %] are converted using the standard Kubelka–Munk algorithm to $A(\lambda)$ extinction spectra (Shimadzu UV-2550 spectrophotometer software).

Glycerol is a well-known effective hyperosmotic agent and is often used for the optical clearing of tissues [13–17,38,39,41,42]. To evaluate the efficiency of optical clearing of ex vivo tissue samples, TTS measurements are usually used, and the efficiency parameter Q is calculated:

$$Q\ (\%) = \{T(t, \lambda) - T(t = 0,\ \lambda)\}/T(t = 0,\ \lambda), \tag{4}$$

where $T(t = 0, \lambda)$ is the transmittance of the tissue sample for a specific wavelength $\lambda$ at the initial time, and $T(t, \lambda)$ is the same at the current time.

The bars on the DRS and TTS charts represent the boundaries of the confidence interval, found as:

$$\sigma = (t_s \mathrm{SD})/(\sqrt{n}) \tag{5}$$

where $t_s$ is Student's coefficient; SD, standard deviation, $n = 5$, $p = 0.95$.

## 4. Results and Discussion

### 4.1. Histological Examination

Histological examination of a tissue sample from the light ovaries revealed cortical and medulla (Figure 5a); an ovarian capsule was also found for the dark ovaries, including the germinal epithelium (single-layered cuboidal epithelium) and the tunica albuginea (subepithelium) (Figure 5b). However, these structural elements are present in both phases.

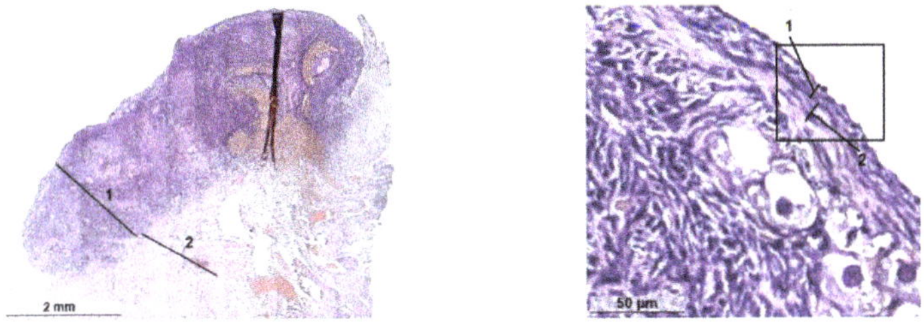


(a) (b)

**Figure 5.** Histology of clinically healthy cat ovaries: (**a**) cortex (1) and medulla (2) in the light ovary (follicular phase); (**b**) the structure of the membrane of the dark ovary (luteal phase): single layer cuboidal epithelium (1) and subepithelial albuginea (2).

In the structure of the ovaries, a clear division into the medullary vascular, fibrous cortical layers is determined. In the latter, numerous groups of primordial follicles are noted subcapsularly (Figure 6a) with nearby primary and secondary follicles. Primordial follicles consist of a primary oocyte surrounded by a single layer of flattened follicular cells. Primary follicles include a larger oocyte and a layer or layers of cuboidal granulosa cells and the shiny sheath of the follicle formed around it (Figure 6b). Secondary follicles have a cavity that separates the oocyte with adjacent granulosa cells (crown radiata) from the layers of granulosa cells lining the follicle from the inside of the basement membrane. Outside of such follicles, layers of theca cells are poorly visualized (Figure 6c). The follicles of the cortical layer are located in hypercellular lean fibrous connective tissue, among which, there are single scar-like structures (Figure 6d).

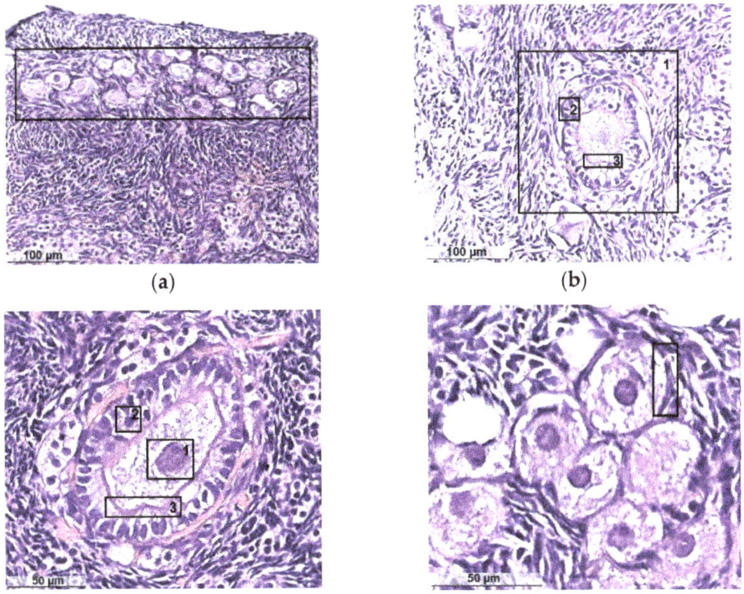


(a) (b)

(c) (d)

**Figure 6.** Histology of clinically healthy cat ovaries in the follicular phase: (**a**) primordial follicles; (**b**) primary follicle (1): granulosa of the follicle (2), shiny sheath of the follicle (3); (**c**) oocyte (1): granulosa of the oocyte (2), the shiny coat of the oocyte (3); (**d**) flattened follicular cells surrounding the oocyte.

In dark specimens, the cortical substance is well developed and dominates over the stroma, and the corpus luteum is well visualized (Figure 7a). Some specimens show corpus luteum hyperplasia without atypia. The stroma is represented by a typical theca tissue without edema. It is moderately developed, the capillary network is multiple and small-focal fresh hemorrhages are noted, in some samples multiple. Vessels are thick-walled and unevenly plethoric in some samples with perivascular fibrosis and hyalinosis. Scattered atretic follicles are visualized (Figure 7b) with internal and external follicular theca with an abundance of blood vessels (Figure 7c) and interstitial connective tissue (Figure 7d).

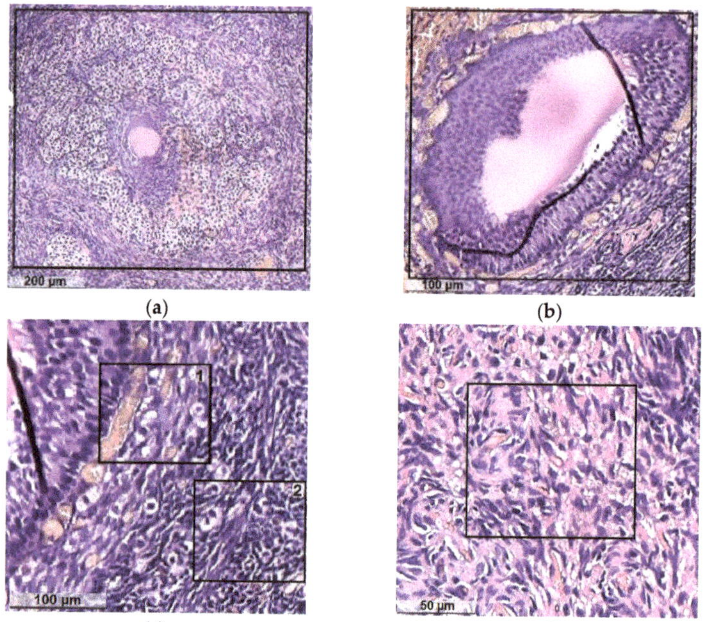


**Figure 7.** Histology of clinically healthy cat ovaries in the luteal phase: (**a**) corpus luteum; (**b**) atretic follicle; (**c**) follicular theca with an abundance of blood vessels: 1—internal, 2—external; (**d**) interstitial connective tissue.

The results of histological studies made it possible to conclude that all the samples of light and dark ovaries taken to study the optical and molecular diffusion properties under the action of glycerol can be attributed to clinically healthy. It was found that light ovaries belong to the follicular phase, and dark ovaries belong to the luteal phase, as the corpus luteum is clearly visualized. Most of the samples showed that the ovaries have a histologically typical structure; in two samples, the histological picture of the ovary with involutive changes was revealed. Necrotic foci, inflammatory infiltrate and atypical growth were not found for all the studied material.

### 4.2. Spectrophotometric Studies

DRSs of the studied samples of cat ovaries in the follicular phase (light ovarian tissue) (F-ph) and in the luteal phase (dark ovarian tissue) (L-ph), initially and after interaction with glycerol, averaged for five samples of each phase of cat ovaries, are shown in Figure 8a,b. It can be seen that the DRSs of both types of samples are almost identical both before and after the diffusion of glycerol. In the UV range, the initial DRSs of the ovarian samples have obvious dips characteristic of the absorption bands of amino acid residues of connective tissue proteins in the form of collagen and reticular fibers, hemoglobin, and porphyrins. In the region of about 415–420 nm and 540–580 nm, the observed dips correspond to the absorption bands of oxyhemoglobin (415, 542 and 576 nm). Water absorption in the measured range of 200–800 nm is insignificant. The main absorption bands in the UV range for common tissue components are located at: 200 nm (proteins), 260 nm (DNA and RNA) and 375 nm (Hb) [16,39]. As a result, tissues are very opaque in the UV range due to the absorption and very strong scattering of light.

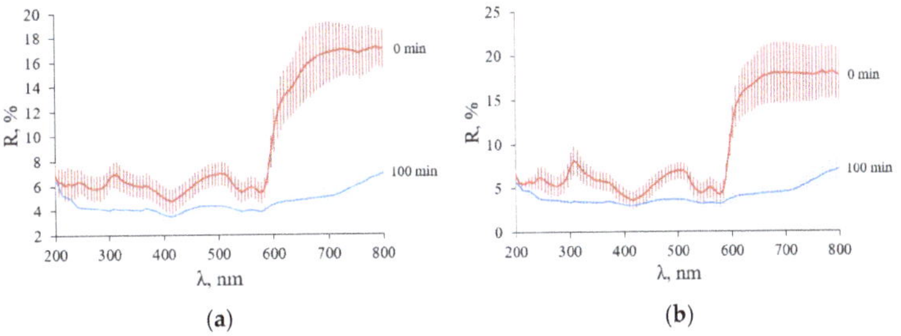


**Figure 8.** DRC spectra in the range from 200 to 800 nm of cat ovary tissue before and after immersion in 99.5% glycerol during 100 min: (**a**) F-ph (light ovarian tissue); (**b**) L-ph (dark ovarian tissue).

The diffusion coefficient of glycerol/interstitial water in the samples was determined from a least squares analysis of a section of the experimental curve characterizing the change in optical density from the time of glycerol action at selected wavelengths. Figure 9a shows the kinetics of DRS during glycerol interaction for 100 min for one of the light samples of cat ovaries. Calculations for each sample were performed for three wavelengths at 600, 700 and 800 nm (Figure 9b).

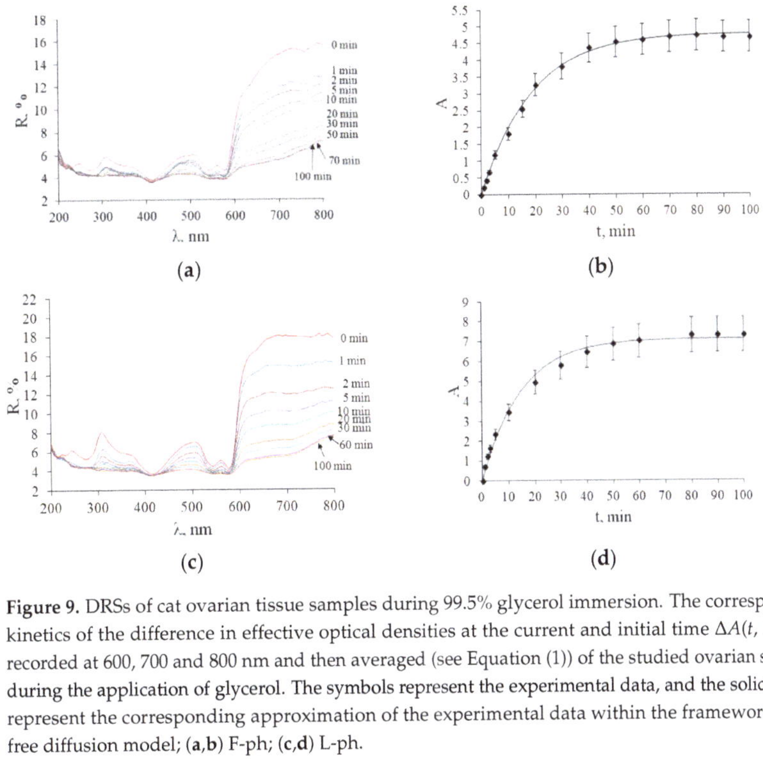


**Figure 9.** DRSs of cat ovarian tissue samples during 99.5% glycerol immersion. The corresponding kinetics of the difference in effective optical densities at the current and initial time $\Delta A(t, \lambda)$ were recorded at 600, 700 and 800 nm and then averaged (see Equation (1)) of the studied ovarian samples during the application of glycerol. The symbols represent the experimental data, and the solid curves represent the corresponding approximation of the experimental data within the framework of the free diffusion model; (**a**,**b**) F-ph; (**c**,**d**) L-ph.

Figure 9c shows the kinetics of DRSs with glycerol action for 100 min on one of the dark cat ovary samples. In both types of tissues (F-ph) and (L-ph) under study, the slowing down and termination of the diffusion process occurred within about 30 min. It can be seen that the interaction of glycerol with the samples leads to a gradual decrease in the reflectance over the entire wavelength range under study. As highly concentrated glycerol was used, it can be assumed that the main outflow of interstitial water from the sample and, consequently, the dehydration of tissues due to the release of water from the sample determine the temporal behavior of the DRS, which indicates a decrease in light scattering and, accordingly, makes it possible to unambiguously relate the rate of diffusion of water molecules to the rate of change in the DRS.

Using Equation (1), we find $\tau$ (diffusion time), which was 22.3 ± 0.6 min for a light sample of ovarian tissue (F-ph), and 17.7 ± 0.7 min for a dark sample of ovarian tissue (L-ph). The average diffusion coefficient for ovarian samples ($n$ = 5) in F-ph was $D = (1.9 \pm 0.2) \cdot 10^{-6}$ cm$^2$/s, and in the L-ph $D = (2.4 \pm 0.2) \cdot 10^{-6}$ cm$^2$/s. The diffusion coefficients of glycerol/interstitial water fluxes in the studied samples determined from the experimental data (Figure 7) according to Equations (1)–(3) and the least squares method are presented in Table 1.

**Table 1.** Kinetic parameters of molecular diffusion in the sections of cat ovaries of the initial thickness $l = 0.8 \pm 0.1$ mm and whole ovary (dehydration time $t_{deh}$ calculated using diffusion coefficient) at the application of 99.5% glycerol.

| Ovarian Tissue | $\tau$, min (Section) | $D \times 10^6$, cm$^2$/s | Tortuosity | $t_{deh}$, h (Whole Ovary) |
|---|---|---|---|---|
| Light (F-ph) | 22.3 ± 0.6 | 1.9 ± 0.2 | 3.9 | 2.3 ± 0.1 |
| Dark (L-ph) | 17.7 ± 0.7 | 2.4 ± 0.2 | 3.5 | 2.1 ± 0.1 |

The data obtained can be compared with the values of the diffusion coefficients of molecular flows, which are caused by the topical application of highly concentrated glycerol. The molecular diffusion coefficient measured in human gingival tissue at the action of 99.5% glycerol was found as $(1.78 \pm 0.22) \times 10^{-6}$ cm$^2$/s ($n$ = 5; $l = 0.59 \pm 0.06$ mm) [41,42], which correlates well with the data received in this paper (Table 1) and the literature data for other tissues [13–17,38–42] taking into account the structural features of tissues and mostly related to tissue water diffusion due to osmotic pressure. As we assume that under the influence of glycerol mainly water migrates in the tissue, the upper limit for the diffusion coefficient should be the rate of water diffusion in the tissue. Based on data for water self-diffusion ($D_W = 3 \times 10^{-5}$ cm$^2$/s [43]), and considering that soft tissues contain up to 75% water, we can estimate the rate of water diffusion in a typical tissue, taking into account the effect of hidden diffusion, which is quantified by the ratio of the path length of the molecular flow between two points in a tissue to the direct distance between these points, named tortuosity [17,44]:

$$\text{Tortuosity} = \sqrt{D_w / D} \tag{6}$$

The tortuosity was estimated at 3.9 for the gingival lamina propria (LP) layer [44] and at 3.5 for the skin dermis [17], which allows to obtain $D_{LP} = 1.9 \times 10^{-6}$ cm$^2$/s and $D_{dermis} = 2.4 \times 10^{-6}$ cm$^2$/s that is in excellent agreement with the measured values of the diffusion coefficient for light ovarian (F-ph) and dark ovarian (L-ph) tissues, with the tortuosity of 3.9 and 3.5, respectively.

Of course, in reality, there is not one flux, but two opposite fluxes: water flux from the tissue and glycerol flux into the tissue. However, at a high concentration of glycerol, the rate of its diffusion in water is low [45]. The diffusion coefficient of glycerol in water at its mass fraction of 84% at room temperature is $2 \times 10^{-7}$ cm$^2$/s, which is an order of magnitude lower than the diffusion rate that we obtained experimentally in this study and which is in good agreement with the water diffusion model in the tissue accounting for the phenomenon of tortuosity.

The diffusion time ($\tau$, min) found from the experimental data for the dark ovary in L-ph turned out to be shorter than for the F-ph (light) samples. This may be because L-ph ovaries contain a larger network of capillaries and are therefore more porous and permeable to migrating molecules. Taking into account the thickness of the whole ovary of $l = 5$ mm and using experimentally determined diffusion coefficients from Table 1 and equation (3), we calculated the dehydration time $t_{deh}$ of the whole ovary under the action of highly concentrated glycerol. The ovary in the luteal phase is dehydrated after ($2.1 \pm 0.1$) hours, and in the follicular phase, a little longer—after ($2.3 \pm 0.1$) hours.

The TTS kinetics for typical samples is shown in Figure 10. In contrast to the DRS, the TTS of the two types of ovarian tissue samples have noticeable differences. The transmittance of both types of samples in the UV is close to zero. The TTSs of light samples (F-ph) show absorption bands of blood hemoglobin, which correlate with the DRSs (Figure 9a). For the TTSs of the dark samples (L-ph), the transmittance is practically zero from 200 to 450 nm and then at 540–590 nm. Obviously, this is due to the fact that this type of ovarian tissue is largely supplied with a capillary network filled with blood. The initial (0 min) and final (100 min) average TTS for all five samples for each type of ovary are shown in Figure 11.

After the complete immersion of the samples, it can be seen that the optical clearing of tissues occurred with the formation of transparency windows. In samples of ovaries in F-ph, the formation of two transparency windows is observed: one in the UV region with a center 350 nm wide ($46 \pm 5$) nm and with a center 500 nm wide ($25 \pm 7$) nm (Figure 11a,b). In the ovary sample in L-ph, the formation of one transparency window in the visible region of the spectrum with a center of 500 and a width of ($21 \pm 6$) nm is observed, and the UV region does not become more transparent (Figure 11c,d).

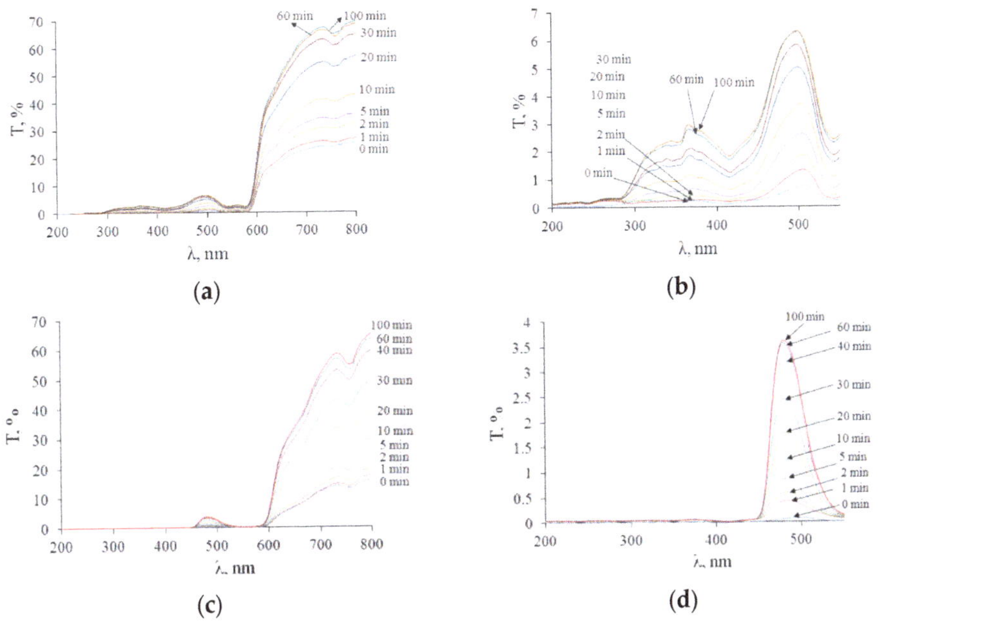


**Figure 10.** TTSs of cat ovarian tissue samples during 99.5% glycerol immersion: (**a**,**b**) F-ph; (**c**,**d**) L-ph.

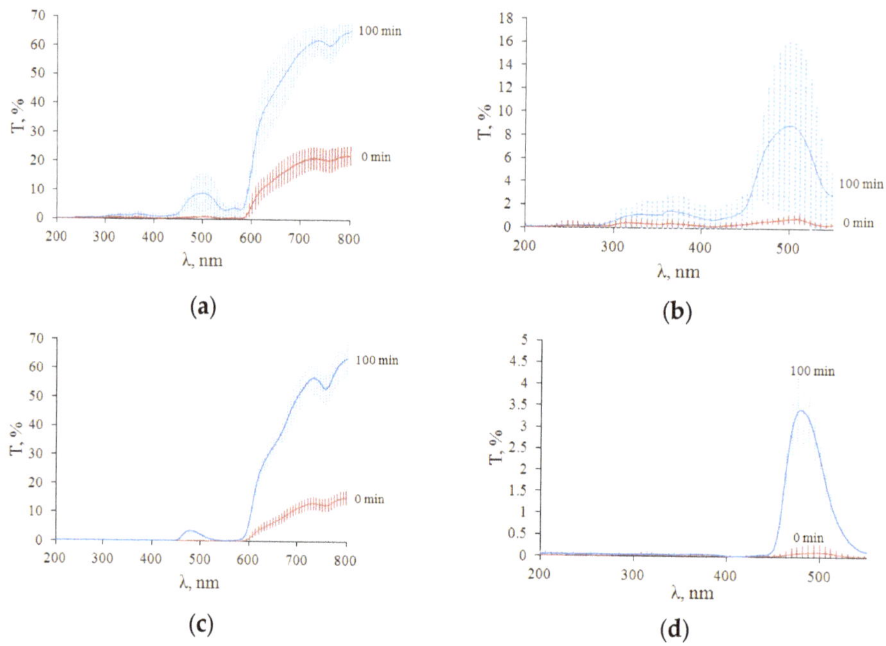


**Figure 11.** Averaged TTS of 5 samples of cat ovarian tissue sections when immersed in 99.5% glycerol: (**a**,**b**) F-ph; (**c**,**d**) L-ph.

The method of immersion optical clearing using hyperosmotic agents, in particular, highly concentrated glycerol, is based on the following mechanisms for suppressing light scattering in tissues. Glycerol induces a partial exchange of tissue water in the interstitial fluid and in the cell cytoplasm and causes tissue dehydration, which, in turn, leads to the matching of the refractive indices of scatterers with the environment (interstitial fluid) and their better packing [13–17,38,39,41,42]. Glycerol has a higher refractive index than interstitial fluid, so when it penetrates the tissues, it also provides refractive index matching, which also causes a decrease in light scattering. As the concentration of glycerol in the tissue becomes sufficiently high, a third mechanism associated with protein dissociation arises [46,47]. However, it is well known that all these mechanisms are reversible and are important for different stages of the optical clearing process [13–17,44,46–50].

The kinetics of change in transmittance for different wavelengths are shown in Figure 12. When a more blood-filled ovary in the luteal phase interacts with glycerol, hemoglobin is rapidly washed out, which goes quickly, as the erythrocytes burst due to osmosis and, together with tissue water, hemoglobin goes into a larger volume of the surrounding solution. Thus, transmission is increased not only by decreasing scattering but also by decreasing the absorption of hemoglobin and its forms, which is a much faster process (see Figure 12d) [50].

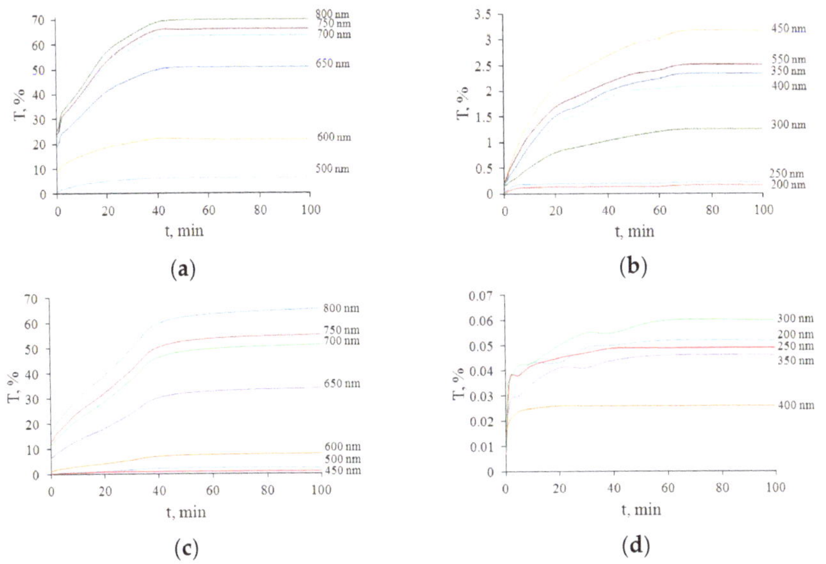


**Figure 12.** The kinetics of the total optical transmittance $T(t)$ of cat ovarian tissue samples for different wavelengths λ when exposed to 99.5% glycerol: (**a**,**b**) F-ph; (**c**,**d**) L-ph.

The efficiency of optical clearing under the influence of 99.5% glycerol ($Q$, %), determined by Equation (4), was calculated using experimental data presented in Figure 9a–d for ovarian tissue in different phases of the cycle (Table 2).

**Table 2.** Efficiency (Q, %) of optical clearing of the ovarian tissue in different phases of the cycle.

| Phase | λ, nm | 200 | 250 | 300 | 350 | 400 | 450 | 500 | 550 | 600 | 650 | 700 | 750 | 800 |
|---|---|---|---|---|---|---|---|---|---|---|---|---|---|---|
| F-ph | *T* (0 min), % | 0.04 | 0.13 | 0.22 | 0.25 | 0.28 | 0.26 | 0.85 | 0.45 | 3.23 | 14.51 | 19.70 | 20.62 | 22.30 |
| | *T* (100 min), % | 0.10 | 0.22 | 0.64 | 1.17 | 0.94 | 1.95 | 8.89 | 3.15 | 13.13 | 45.50 | 57.81 | 60.82 | 65.50 |
| | Q, % | 148 | 69 | 190 | 370 | 236 | 650 | 946 | 600 | 306 | 213 | 193 | 195 | 194 |
| L-ph | *T* (0 min), % | 0.009 | 0.013 | 0.018 | 0.021 | 0.004 | 0.021 | 0.11 | 0.02 | 1.19 | 6.50 | 11.32 | 12.61 | 15.21 |
| | *T* (100 min), % | 0.046 | 0.047 | 0.044 | 0.046 | 0.012 | 0.074 | 2.39 | 0.11 | 7.50 | 32.91 | 49.54 | 53.71 | 63.60 |
| | Q, % | 411 | 261 | 144 | 119 | 200 | 250 | 2074 | 450 | 529 | 406 | 338 | 326 | 318 |

Figure 13 shows visual changes in the studied tissue samples before and after optical clearing with 99.5% glycerol. The images were taken using the camera of a Samsung Galaxy A51 smartphone with a resolution of 48 MP. To obtain the photos, the samples were placed on a sheet of white paper with white light falling from above.

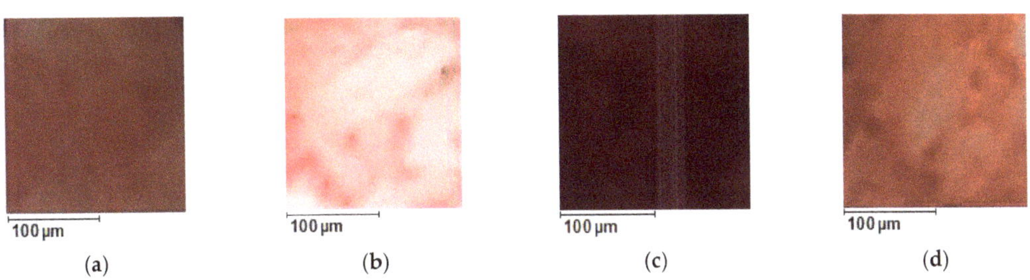


**Figure 13.** Photos of the area of the samples of the cat ovaries, magnified 50 times: F-ph (**a**) before optical clearing; (**b**) after optical clearing; L-ph (**c**) before optical clearing; (**d**) after optical clearing. The optical clearing agent is 99.5% glycerol.

In the UV range, the efficiency of optical clearing of cat ovarian tissue with 99.5%-glycerol is high and reaches 370% for the F-ph samples and 411% for the L-ph samples. The absolute optical transmittance is not high and reaches 3.5% at 350 nm for F-ph samples (Figure 10b) and only 0.2% for L-ph samples because of strong light scattering combined with strong absorption by the endogenous chromophores, including blood hemoglobin. At 500 nm, the optical clearing efficiency reaches 946% for the F-ph samples and 2074% for the L-ph samples, with the total transmittance up to 16% for the F-ph samples and 4% for the L-ph samples. At 600 nm, the optical clearing efficiency reaches 306% for the F-ph samples and 529% for the L-ph samples, with the total transmittance up to 5% for the F-ph samples and 0.3% for the L-ph samples.

In the so-called "first therapeutic/diagnostic window" at 650–800 nm [16], the efficiency of optical clearing is not the highest and reaches 213% for the F-ph samples and 405% for the L-ph samples. However, due to the absence of strong absorption bands of endogenous chromophores in this region, the absolute transmittance is quite large and amounts to 70% (Figure 10a,c).

Similar results were obtained when using highly concentrated glycerol for the optical clearing of colorectal tissues in normal conditions and in polyposis pathologies, as well as healthy gingival tissue [44]. For the colonic mucosa, two windows of dynamic transparency were identified in the UV range from 200 to 260 nm and from 260 to 418 nm, and a lower efficiency of optical clearing was shown in the long-wave visible/NIR region with a high level of absolute transmittance.

## 5. Conclusions

Two groups of cat ovarian samples were studied. The histological examination of these samples revealed a difference between these groups. It was determined that the light ovaries are in the follicular phase and do not contain a corpus luteum. In the darker ovaries, corpora luteal of various stages were found, which corresponds to the luteal phase of the cycle. The diffuse reflectance and total transmittance of samples in the pre-luteal and luteal phases of the cycle were determined by diffuse spectroscopy. Using the optical kinetics of ovarian tissue samples at glycerol action, glycerol/tissue water diffusion coefficient was estimated, $D = (1.9 \pm 0.2) \times 10^{-6}$ cm$^2$/s for ovaries in the follicular stage of the cycle and $D = (2.4 \pm 0.2) \times 10^{-6}$ cm$^2$/s for ovaries in the luteal phase of the cycle. Using the obtained diffusion coefficients, it was possible to obtain the time for the complete dehydration of the whole ovary at glycerol action. The time of the complete dehydration of the ovary sections 0.8 mm-thick in the follicular phase was estimated as 22.3 min, and in the luteal phase, 17.7 min. These data can be used to evaluate total ovarian dehydration at concentrated glycerol applications. In general, the data received in this study can be used for designing the protocols for drug delivery and the cryopreservation of organs.

The total optical transmittance of the ovaries in the follicular phase is much higher than in the luteal phase, which is associated with an anastomosis of an extensive network of capillaries and abundant blood supply to the ovary during this phase of the cycle. Thus, using diffuse spectroscopy, it is possible to fix a fairly short period of formation of the corpus luteum, which is an extremely dynamic temporary organ—a gland that produces progesterone and plays a central role in the reproductive process. The emergence and development of the corpus luteum are extremely rapid with a high cell turnover and a strong blood supply that is primarily regulated by angiogenic growth factors. This ability to accurately determine the timing of lutein formation is extremely important in the study of infertility of unknown origin and when using assisted reproductive technologies.

Optical clearing technology using hyperosmotic agents, in particular glycerol, reduces light scattering and, as a result, improves the penetration depth of light. When ovarian tissue was immersed in glycerol, the efficiency of optical clearing reached 370% in the wavelength range from 280 to 410 nm and up to 946% in the range of 430–550 nm. This effect can be used in therapeutic and diagnostic clinical applications to study molecular structures deep in the tissue.

The optical clearing technology presented in this study also improves tissue transparency in the UV range and may be useful for the effective application of existing and future UV biomedical spectroscopies and therapies, in particular, to study the structure and dynamics of proteins using UV resonance Raman spectroscopy [51], for the general use of deep UV Raman spectroscopy [52], the detection of pathologies such as gliomas with UV fluorescence excitation [53], or the use of deep UV fluorescence microscopy in cell biology and tissue histology [54] and for other biomedical optical technologies, where UV excitation is fundamentally important.

The optical clearing and diffusion–kinetic properties of a number of other, more common cryopreservatives, such as DMSO, ethylene glycol and PrOH, have been studied for muscle and skin tissues [13–16,48]. It seems important to carry out similar quantitative studies for animal ovaries in the follicular and luteal phases of the cycle.

The studied glycerol-induced perfusion kinetics of ovarian tissues is of great importance both for the development of clinical protocols for optical tissue clearing in laparoscopic diagnostic or surgical applications and for the cryopreservation of ovarian tissues. Moreover, the technology can potentially be used for the optical monitoring of changes in tissue structure during the storage of a cryopreserved organ.

An in-depth interdisciplinary study is needed to reduce side effects and preserve fertility in women with cancer [3–6,10], including those complicated by diabetes mellitus [55], using new technologies for ovarian cryopreservation, including surgical procedures for ovarian transplantation and new reproductive technologies. The successful cryopreservation and subsequent thawing of the transplanted ovary largely depend on the knowledge of the quantitative characteristics of the perfusion–kinetic processes during the freezing and thawing of the organ.

**Author Contributions:** Conceptualization, V.V.T.; Methodology, A.A.S.; Formal analysis, A.A.S. and V.V.T.; Investigation A.A.S. and A.S.R.; Resources, A.A.S. and V.V.T.; Writing—original draft preparation, A.A.S.; Writing—review and editing, V.V.T.; Project administration, A.A.S. and V.V.T.; Funding acquisition, A.A.S. and V.V.T. All authors have read and agreed to the published version of the manuscript.

**Funding:** Russian Science Foundation Grant No. 22-75-00021 (A.A.S.); Ministry of Science and Higher Education of the Russian Federation within the framework of a state assignment (project No. FSRR-2023-0007) (V.V.T.).

**Institutional Review Board Statement:** The study was conducted in accordance with the Declaration of Helsinki, and approved by the Ethics Committee of Saratov State Medical University named after V. I. Razumovsky (No. 4 dated 1 November 2022).

**Informed Consent Statement:** Not applicable.

**Data Availability Statement:** Data are available from the authors.

**Acknowledgments:** A.A.S. was supported by the Russian Science Foundation Grant No. 22-75-00021. V.V.T. was supported by the Ministry of Science and Higher Education of the Russian Federation within the framework of a state assignment (project No. FSRR-2023-0007).

**Conflicts of Interest:** The authors declare no conflict of interest. The funders had no role in the design of the study; in the collection, analyses, or interpretation of data; in the writing of the manuscript; or in the decision to publish the results.

## References

1. Del-Pozo-Lerida, S.; Salvador, C.; Martínez-Soler, F.; Tortosa, A.; Perucho, M.; Gimenez-Bonaf, P. Preservation of fertility in patients with cancer (Review). *Oncol. Rep.* **2019**, *41*, 2607–2614. [CrossRef] [PubMed]
2. Siegel, R.L.; Miller, K.D.; Fuchs, H.E.; Jemal, A. Cancer statistics. *CA Cancer J. Clin.* **2022**, *72*, 7–33. [CrossRef]
3. Taylan, E.; Oktay, K. Fertility preservation in gynecologic cancers. *Gynecol. Oncol.* **2019**, *155*, 522–529. [CrossRef] [PubMed]
4. Medrano, J.V.; Andrés, M.D.M.; García, S.; Herraiz, S.; Vilanova-Pérez, T.; Goossens, E.; Pellicer, A. Basic and clinical approaches for fertility preservation and restoration in cancer patients. *Trends Biotechnol.* **2018**, *36*, 199–215. [CrossRef] [PubMed]

5. Armstrong, A.G.; Kimler, B.F.; Smith, B.M.; Woodruff, T.K.; Pavone, M.E.; Duncan, F.E. Ovarian tissue cryopreservation in young females through the Oncofertility Consortium's National Physicians Cooperative. *Future Oncol.* **2018**, *14*, 363–378. [CrossRef]
6. Santos, M.L.; Pais, A.S.; Almeida Santos, T. Fertility preservation in ovarian cancer patients. *Gynecol. Endocrinol.* **2021**, *37*, 483–489. [CrossRef]
7. Hinkle, K.; Orwig, K.E.; Valli-Pulaski, H.; Taylor, S.; Leeuwen, K.; Carpentieri, D.; Walsh, A. Cryopreservation of ovarian tissue for pediatric fertility. *Biopreserv. Biobank.* **2021**, *19*, 130–135. [CrossRef]
8. Rivas Leonel, E.C.; Lucci, C.M.; Amorim, C.A. Cryopreservation of human ovarian tissue: A review. *Transfus. Med. Hemother.* **2019**, *46*, 173–181. [CrossRef]
9. Gjeterud, J.; Kristensen, S.G.; Fedder, J. Indications for cryopreservation and autotransplantation of ovarian tissue. *Tidsskr. Nor. Laegeforen.* **2021**, *141*, 2015–2021. [CrossRef]
10. Del Valle, L.; Corchon, S.; Palop, J.; Rubio, J.M.; Celda, L. The experience of female oncological patients and fertility preservation: A phenomenology study. *Eur. J. Cancer Care* **2022**, *31*, e13757. [CrossRef]
11. Laguerre, M.D.; Arkerson, B.J.; Robinson, M.A.; Moawad, N.S. Outcomes of laparoscopic management of chronic pelvic pain and endometriosis. *J. Obstet. Gynecol.* **2022**, *42*, 146–152. [CrossRef] [PubMed]
12. Vignarajan, C.P.; Malhotra, N.; Singh, N. Ovarian reserve and assisted reproductive technique outcomes after laparoscopic proximal tubal occlusion or salpingectomy in women with hydrosalpinx undergoing in vitro fertilization: A randomized controlled trial. *J. Minim. Invasive Gynecol.* **2019**, *26*, 1070–1075. [CrossRef] [PubMed]
13. Tuchin, V.V. *Optical Clearing of Tissues and Blood*; SPIE Press: Bellingham, WA, USA, 2006.
14. Oliveira, L.M.C.; Tuchin, V.V. *The Optical Clearing Method—A New Tool for Clinical Practice and Biomedical Engineering*; Springer: Cham, Switzerland, 2019.
15. Tuchin, V.V.; Zhu, D.; Genina, E.A. *Handbook of Tissue Optical Clearing: New Prospects in Optical Imaging*; CRC Press: Boca Raton, FL, USA, 2022.
16. Tuchin, V.V. *Tissue Optics: Light Scattering Methods and Instruments for Medical Diagnostics*, 3rd ed.; SPIE Press: Bellingham, WA, USA, 2015.
17. Tuchina, D.K.; Meerovich, I.G.; Sindeeva, O.A.; Zherdeva, V.V.; Savitsky, A.P.; Bogdanov, A.A., Jr.; Tuchin, V.V. Magnetic resonance contrast agents in optical clearing: Prospects for multimodal tissue imaging. *J. Biophotonics* **2020**, *13*, e201960249. [CrossRef] [PubMed]
18. Sethi, A.; Singh, N.; Patel, G. Is Hysterosalpingography using Magnetic Resonance Imaging a promising tool in infertility patients? *JBRA Assist. Reprod.* **2022**, *26*, 161. [CrossRef] [PubMed]
19. Arapaki, A.; Christopoulos, P.; Kalampokas, E.; Triantafyllidou, O.; Matsas, A.; Vlahos, N.F. Ovarian tissue cryopreservation in children and adolescents. *Children* **2022**, *9*, 1256. [CrossRef]
20. Shapira, M.; Dolmans, M.M.; Silber, S.; Meirow, D. Evaluation of ovarian tissue transplantation: Results from three clinical centers. *Fertil. Steril.* **2020**, *114*, 388–397. [CrossRef] [PubMed]
21. Donnez, J.; Dolmans, M.M.; Demylle, D.; Jadoul, P.; Pirard, C.; Squifflet, J.; Martinez-Madrid, B.; van Langendonckt, A. Livebirth after orthotopic transplantation of cryopreserved ovarian tissue. *Lancet* **2004**, *364*, 1405–1410. [CrossRef]
22. Von Wolff, M.; Donnez, J.; Hovatta, O.; Keros, V.; Maltaris, T.; Montag, M.; Salle, B.; Sonmezer, M.; Andersen, C.Y. Cryopreservation and autotransplantation of human ovarian tissue prior to cytotoxic therapy—A technique in its infancy but already successful in fertility preservation. *Eur. J. Cancer* **2009**, *45*, 1547–1553. [CrossRef]
23. Shi, Q.; Xie, Y.; Wang, Y.; Li, S. Vitrification versus slow freezing for human ovarian tissuse cryopreservation: A systematic review and meta-anlaysis. *Sci. Rep.* **2017**, *7*, 8538–8547. [CrossRef]
24. Lee, S.; Ozkavukcu, S.; Ku, S.Y. Current and Future Perspectives for Improving Ovarian Tissue Cryopreservation and Transplantation Outcomes for Cancer Patients. *Reprod. Sci.* **2021**, *28*, 1746–1758. [CrossRef]
25. Kometas, M.; Christman, G.M.; Kramer, J.; Rhoton-Vlasak, A. Methods of Ovarian Tissue Cryopreservation: Is Vitrification Superior to Slow Freezing? -Ovarian Tissue Freezing Methods. *Reprod. Sci.* **2021**, *28*, 3291–3302. [CrossRef] [PubMed]
26. Amorim, C.A.; David, A.; Van Langendonckt, A.; Dolmans, M.M.; Donnez, J. Vitrification of human ovarian tissue: Effect of different solutions and procedures. *Fertil. Steril.* **2011**, *95*, 1094–1097. [CrossRef] [PubMed]
27. Klocke, S.; Bündgen, N.; Köster, F.; Eichenlaub-Ritter, U.; Griesinger, G. Slow-freezing versus vitrification for human ovarian tissue cryopreservation. *Arch. Gynecol. Obstet.* **2015**, *291*, 419–426. [CrossRef] [PubMed]
28. Oktay, K.; Karlikaya, G. Ovarian function after transplantation of frozen, banked autologous ovarian tissue. *N. Engl. J. Med.* **2000**, *342*, 1919–1924. [CrossRef] [PubMed]
29. Bedaiwy, M.A.; Falcone, T. Harvesting and autotransplantation of vascularized ovarian grafts: Approaches and techniques. *Reprod. Biomed.* **2007**, *14*, 360–371. [CrossRef]
30. Martinez-Madrid, B.; Dolmans, M.M.; Van Langendonckt, A.; Defrere, S.; Donnez, J. Freeze-thawing intact human ovary with its vascular pedicle with a passive cooling device. *Fertil. Steril.* **2004**, *82*, 1390–1394. [CrossRef]
31. Burkman, R.T. Atlas of pelvic anatomy and gynecologic surgery. *JAMA J. Am. Med. Assoc.* **2011**, *306*, 2031–2038. [CrossRef]
32. Worne, F. Color atlas of veterinary anatomy. Volume 3: The dog and cat. *Aust. Vet. J.* **2011**, *89*, 124–136. [CrossRef]
33. Yada, H.; Hosokawa, K.; Tajima, K.; Hasegawa, Y.; Kotsuji, F. Role of ovarian theca and granulosa cell interaction in hormone production and cell growth during the bovine follicular maturation process. *Biol. Reprod.* **1999**, *61*, 1480–1486. [CrossRef]
34. Lamond, D.R.; Drost, M. Blood supply to the bovine ovary. *J. Anim. Sci.* **1974**, *38*, 106–112. [CrossRef]

35. Smith, M.F.; McIntush, E.W.; Smith, G.W. Mechanisms associated with corpus luteum development. *J. Anim. Sci.* **1994**, *72*, 1857–1872. [CrossRef] [PubMed]
36. Reynolds, L.P.; Redmer, D.A. Growth and development of the corpus luteum. *J. Reprod. Fertil. Suppl.* **1999**, *54*, 181–191. [CrossRef] [PubMed]
37. Refractiveindex.info Database. Available online: https://refractiveindex.info/?shelf=organic&book=glycerol&page=Birkhoff (accessed on 23 January 2023).
38. Genina, E.A.; Bashkatov, A.N.; Tuchin, V.V. Tissue optical immersion clearing. *Expert Rev. Med. Devices* **2010**, *6*, 825–842. [CrossRef] [PubMed]
39. Martins, I.S.; Silva, H.F.; Lazareva, E.N.; Chernomyrdin, N.V.; Zaytsev, K.I.; Oliveira, L.M.; Tuchin, V.V. Measurement of tissue optical properties in a wide spectral range: A review. *Biomed. Opt. Express* **2023**, *14*, 249–298. [CrossRef]
40. Kotyk, A.; Janacek, K. *Cell Membrane Transport. Principles and Techniques*, 2nd ed.; Plenum Press: New York, NY, USA; London, UK, 1974; pp. 501–583.
41. Selifonov, A.A.; Tuchin, V.V. Chapter 36. Optical Clearing and Molecular Diffusivity of Hard and Soft Oral Tissues. In *Handbook of Tissue Optical Clearing: New Prospects in Optical Imaging*; Tuchin, V.V., Zhu, D., Genina, E.A., Eds.; CRC Press: Boca Raton, FL, USA, 2022; pp. 629–646. [CrossRef]
42. Selifonov, A.; Tuchin, V.V. Determination of the kinetic parameters of glycerol diffusion in the gingival and dentinal tissue of a human tooth using optical method: In vitro studies. *Opt. Quantum Electron.* **2020**, *52*, 123–127. [CrossRef]
43. Schwindt, D.A.; Wilhelm, K.P.; Maibach, H.I. Water diffusion characteristics of human stratum corneum at different anatomical sites in vivo. *J. Investig. Dermatol.* **1998**, *111*, 385–389. [CrossRef]
44. Carneiro, I.; Carvalho, S.; Henrique, R.; Selifonov, A.; Oliveira, L.; Tuchin, V.V. Enhanced ultraviolet spectroscopy by optical clearing for biomedical applications. *IEEE J. Sel. Top. Quantum Electron.* **2021**, *27*, 7200108. [CrossRef]
45. Akinkunmi, F.O.; Jahn, D.A.; Giovambattista, N. Effects of temperature on the thermodynamic and dynamical properties of glycerol−water mixtures: A computer simulation study of three different force fields. *J. Phys. Chem. B* **2015**, *119*, 6250–6261. [CrossRef]
46. Yeh, A.; Choi, B.; Nelson, J.S.; Tromber, B.J. Reversible dissociation of collagen in tissues. *J. Investig. Dermatol.* **2003**, *121*, 1332–1335. [CrossRef]
47. Hirshburg, J.; Choi, B.; Nelson, J.S.; Yeh, A.T. Collagen solubility correlates with skin optical clearing. *J. Biomed. Opt.* **2006**, *11*, 040501. [CrossRef]
48. Oliveira, L.; Carvalho, M.I.; Nogueira, E.; Tuchin, V.V. Diffusion characteristics of ethylene glycol in skeletal muscle. *J. Biomed. Opt.* **2015**, *20*, 051019. [CrossRef] [PubMed]
49. Carneiro, I.; Carvalho, S.; Henrique, R.; Oliveira, L.; Tuchin, V.V. A robust ex vivo method to evaluate the diffusion properties of agents in biological tissues. *J. Biophotonics* **2019**, *12*, e201800333. [CrossRef] [PubMed]
50. Oliveira, L.R.; Ferreira, R.M.; Pinheiro, M.R.; Silva, H.F.; Tuchin, V.V.; Oliveira, L.M. Broadband spectral verification of optical clearing reversibility in lung tissue. *J. Biophotonics* **2022**, *16*, e202200185. [CrossRef] [PubMed]
51. Jakubek, R.S.; Handen, J.; White, S.E.; Asher, S.A.; Lednev, I.K. Ultraviolet resonance Raman spectroscopic markers for protein structure and dynamics. *TrAC Trends Anal. Chem.* **2018**, *103*, 223–229. [CrossRef]
52. Arnold, B.R.; Cooper, C.E.; Matrona, M.R.; Emge, D.K.; Oleske, J.B. Stand-off deep-UV Raman spectroscopy. *Can. J. Chem.* **2018**, *96*, 614–620. [CrossRef]
53. Mehidine, H.; Chalumeau, A.; Poulon, F.; Jamme, F.; Varlet, P.; Devaux, B.; Refregiers, M.; Haidar, D.A. Optical signatures derived from deep UV to NIR excitation discriminates healthy samples from low and high grades glioma. *Sci. Rep.* **2019**, *9*, 8786–8792. [CrossRef]
54. Jamme, F.; Kascakova, S.; Villette, S.; Allouche, F.; Pallu, S.; Rouam, V.; Refregiers, M. Deep UV autofluorescence microscopy for cell biology and tissue histology. *Biol. Cell* **2013**, *105*, 277–288. [CrossRef]
55. Saed, L.; Varse, F.; Baradaran, H.R.; Moradi, Y.; Khateri, S.; Friberg, E.; Khazaei, Z.; Gharahjeh, S.; Tehrani, S.; Sioofy-Khojine, A.B.; et al. The effect of diabetes on the risk of endometrial cancer: An updated a systematic review and meta-analysis. *BMC Cancer* **2019**, *19*, 527. [CrossRef]


**Disclaimer/Publisher's Note:** The statements, opinions and data contained in all publications are solely those of the individual author(s) and contributor(s) and not of MDPI and/or the editor(s). MDPI and/or the editor(s) disclaim responsibility for any injury to people or property resulting from any ideas, methods, instructions or products referred to in the content.

*diagnostics* MDPI

*Article*

# Application Prospects of FTIR Spectroscopy and CLSM to Monitor the Drugs Interaction with Bacteria Cells Localized in Macrophages for Diagnosis and Treatment Control of Respiratory Diseases

**Igor D. Zlotnikov [1], Alexander A. Ezhov [2], Maksim A. Vigovskiy [3,4], Olga A. Grigorieva [3,4], Uliana D. Dyachkova [3,4], Natalia G. Belogurova [1] and Elena V. Kudryashova [1,*]**

1 Faculty of Chemistry, Lomonosov Moscow State University, Leninskie Gory, 1/3, 119991 Moscow, Russia
2 Faculty of Physics, Lomonosov Moscow State University, Leninskie Gory, 1/2, 119991 Moscow, Russia
3 Medical Research and Education Center, Institute for Regenerative Medicine, Lomonosov Moscow State University, 27/10, Lomonosovsky Ave., 119192 Moscow, Russia
4 Faculty of Medicine, Lomonosov Moscow State University, 27/1, Lomonosovsky Prosp., 119192 Moscow, Russia
* Correspondence: helena_koudriachova@hotmail.com

**Abstract:** Visualization of the interaction of drugs with biological cells creates new approaches to improving the bioavailability, selectivity, and effectiveness of drugs. The use of CLSM and FTIR spectroscopy to study the interactions of antibacterial drugs with latent bacterial cells localized in macrophages create prospects to solve the problems of multidrug resistance (MDR) and severe cases. Here, the mechanism of rifampicin penetration into *E. coli* bacterial cells was studied by tracking the changes in the characteristic peaks of cell wall components and intracellular proteins. However, the effectiveness of the drug is determined not only by penetration, but also by efflux of the drugs molecules from the bacterial cells. Here, the efflux effect was studied and visualized using FTIR spectroscopy, as well as CLSM imaging. We have shown that because of efflux inhibition, eugenol acting as an adjuvant for rifampicin showed a significant (more than three times) increase in the antibiotic penetration and the maintenance of its intracellular concentration in *E. coli* (up to 72 h in a concentration of more than 2 μg/mL). In addition, optical methods have been applied to study the systems containing bacteria localized inside of macrophages (model of the latent form), where the availability of bacteria for antibiotics is reduced. Polyethylenimine grafted with cyclodextrin carrying trimannoside vector molecules was developed as a drug delivery system for macrophages. Such ligands were absorbed by CD206+ macrophages by 60–70% versus 10–15% for ligands with a non-specific galactose label. Owing to presence of ligands with trimannoside vectors, the increase in antibiotic concentration inside macrophages, and thus, its accumulation into dormant bacteria, is observed. In the future, the developed FTIR+CLSM techniques would be applicable for the diagnosis of bacterial infections and the adjustment of therapy strategies.

**Keywords:** FTIR spectroscopy; CLSM; macrophage; latent infection; drug resistance

**Citation:** Zlotnikov, I.D.; Ezhov, A.A.; Vigovskiy, M.A.; Grigorieva, O.A.; Dyachkova, U.D.; Belogurova, N.G.; Kudryashova, E.V. Application Prospects of FTIR Spectroscopy and CLSM to Monitor the Drugs Interaction with Bacteria Cells Localized in Macrophages for Diagnosis and Treatment Control of Respiratory Diseases. *Diagnostics* **2023**, *13*, 698. https://doi.org/10.3390/diagnostics13040698

Academic Editor: Viktor Dremin

Received: 26 December 2022
Revised: 8 February 2023
Accepted: 9 February 2023
Published: 12 February 2023

**Copyright:** © 2023 by the authors. Licensee MDPI, Basel, Switzerland. This article is an open access article distributed under the terms and conditions of the Creative Commons Attribution (CC BY) license (https://creativecommons.org/licenses/by/4.0/).

## 1. Introduction

Respiratory tract diseases (tuberculosis, pneumonia, mycoplasmosis etc) caused by pathogenic microorganisms are an acute problem in modern society [1,2]. Moreover, the resistant forms of microorganisms practically insensitive to antibiotics are caused by a number of factors, including the drug efflux effect, which is particularly dangerous [3–8]. For several decades, the main causative agents of bacterial forms of respiratory diseases have been *S. pneumoniae*, *M. tuberculosis*, *H. influenzae* type b, *S. pyogenes*, *M. catarrhalis*, and *S. aureus* [9]. These pathogens develop resistance to amoxicillin, rifampicin, macrolides, and cephalosporins. Resistance to antibiotics is multifactorial and may be caused by one or a combination of mutations in target genes, increased production of multidrug-resistant outflow pumps (MDR), or modifying enzymes and/or target-protecting proteins [10,11]. In this paper, special attention is focused on the study of efflux and ligands that increase specificity, which are considered as one of the main processes that reduces the effectiveness of antibiotics.

A special problem in the treatment of diseases are pathogens localized in macrophages and granulomas, thereby passing into a dormant form (but still dangerous) and are difficult to treat, as seen in some forms of tuberculosis, leishmaniasis, and respiratory diseases caused by chlamydial infections, which are particularly prevalent in children and people with weakened immunity, etc. [6,12,13]. This work is devoted to the development of spectral and optical methods for studying drug interactions with bacterial cells in their individual form and inside macrophages—as a model of latent infections. The considered approaches are perspective in application of the bacterial disease diagnosis by analysis of the FTIR spectra of pulmonary lavages using library data on microorganisms, as well as the optimization of the action of drugs for each patient. The possibility of analyzing pulmonary lavage is described in the work [14], where the combination of FTIR and confocal microscopy establishes the composition of the biological fluid [14]. However, for accurate validation, reference spectra and the use of several methods are necessary. The spectra biological objects such as cells are usually too multifactorial and it is not yet fully clear how to specify and isolate analytically significant signals; however, with the development of technologies, new prospects are opening up, such as time-resolved spectroscopy and 2D analysis. The literature describes the use of Raman scattering for the study of bacterial cells and identification of pathogens and strains [15]. Currently, the method is unreliable and, accordingly, not very informative, since it analyzes not the intrinsic spectra of substances in the giant Raman scattering, but only indirect ones—the effect of bacteria on the optical properties of nanoparticles. It strongly depends on the quality of the sample application, on the batch of particles, and on the conditions of strain cultivation [16,17].

On the contrary, FTIR spectroscopy, which is a highly informative method of analyzing chemical compounds, provides information about chemical bonds and the microenvironment of molecules, and is quite sensitive to changes in fine organization at the molecular level [6,18–20]. This makes it possible to study the structures of complex biological objects (cells, organelles), which is of limited use for other spectral methods that require the optical transparency of samples, which clearly does not apply to cells. Practical applications of FTIR spectroscopy include: analysis of biological tissues [21], tumor diagnosis [22], identification of pathogenic bacteria [23], and the study of molecular mechanisms of adaptation to changes in external conditions [24]—which opens up the opportunities to study the development of resistance. We assume that by using FTIR spectroscopy it would be possible to monitor the course of treatment, or the effectiveness of the drug used, as well as analyzing the biological fluids, regulating treatment regimes, and determining sensitivity to antibiotics. The advantages of FTIR spectroscopy include a small amount of substance for analysis (50 μL), non-invasive and giving numerical reagents, and biochemical changes at the molecular level. It has been widely discussed that the optical method of FTIR spectroscopy can potentially be used by doctors to accelerate the diagnosis of a patient or as an auxiliary method of analysis—during or before surgery [25].

Confocal Laser Scanning Microscopy (CLSM) is an optical imaging technique for increasing the optical resolution and contrast of micrography by using a spatial point hole to block out-of-focus light during image formation. Confocal microscopy allows direct, non-invasive sequential optical cutting of intact, thick, living samples with minimal sample preparation. Practical applications of CLSM include imaging of various tissues, cells, and drug interactions. CLSM provides efficient characterization of the physicochemical properties of drug delivery systems [26], diagnostics, and examination of cancerous tissues [27] and visualization of bacteria [6,28–30]. Confocal microscopy makes it possible to visualize the accumulation of drugs in cells and to study the mechanisms affecting the effectiveness of the therapeutic agent [6,28,29,31–34].

As possible promising experimental bases to solve the problem of latent infections localized in macrophages and multidrug resistance of bacteria, we propose two approaches: the use of a targeted drug delivery system to macrophages to concentrate drugs in the lungs [6,28,31,32,35–56], as well as the use of adjuvants (allylbenzenes and terpenoids) that inhibit efflux and increase the permeability of the membrane of pathogenic microorganisms [6–8,34,57–61]. We used rifampicin (Rif) [35,53,62] and doxorubicin (Dox) [35,57] as model (fluorophore) drugs, and studied their synergism with adjuvants (terpenoids from plant extracts) as well as the molecular mechanisms of such combined drug action that can be visualized by optical and spectral methods.

Thus, this study is aimed at developing a new approach based on spectral and optical methods for possible use in medical practice: monitoring the course of treatment based on spectral data of biological fluids, potential diagnosis of bacterial diseases, and strengthening existing therapeutic formulations.

## 2. Materials and Methods

### 2.1. Reagents

Carbonyldiimidazole (CDI) was obtained from GL Biochem Ltd. (Shanghai, China) via an intermediary Himprocess (Moscow, Russia). D-mannose, galactose, PEI 1.8 kDa (branched), fluorescein isothiocyanate (FITC), $NaBH_3CN$, DMF, DMSO, $Et_3N$, 2-hydroxypropyl-β-cyclodextrin (HPCD), 1M 2,4,6-trinitrobenzenesulfonic acid, rifampicin, and doxorubicin were obtained from Sigma Aldrich (St. Louis, MI, USA). Eosin-5-maleimide was purchased from Invitrogen (Molecular Probes, Eugene, Oregon, USA). Mannotriose-di-(N-acetyl-D-glucosamine) was obtained from Dayang Chem Co., Ltd. (Hangzhou, China). Eugenol and safrole at the highest commercial quality were purchased from Acros Organics (Flanders, Belgium). The preparation of apiol and plant extracts was carried out in the same way as described earlier [63]. Other chemicals such as salts and acids were obtained from Reakhim Production (Moscow, Russia).

### 2.2. Synthesis and Characterization of Conjugates

#### 2.2.1. Synthesis of Grafted Chitosan and Cyclodextrin

The synthesis, purification, and characterization of conjugates were carried out as described earlier [6,20,50], including steps of HPCD activation by carbonyldiimidazole, subsequent conjugation with PEI amino groups, and modification by three types of carbohydrate labels: linear galactose, linear mannose, and trimannoside—to determine the affinity to CD206 receptors of macrophages. Introduction of the FITC label: to the aqueous solution of PEI1.8 (5% in 0.01M HCl, 1 g), a solution of FITC (15 mg in 1.5 mL DMSO) was added drop by drop with stirring; the pH was brought to 9.2 (sodium borate buffer, 0.1M). The mixture was incubated at 40 °C for 1 h, followed by purification by dialysis against water (cut-off 1 kDa) for 6 h. Purification by dialysis and HPLC: characterization was performed using by NMR, FTIR spectroscopy, analysis of nanoparticle trajectories, and dynamic light scattering.

#### 2.2.2. Dynamic Light Scattering (DLS)

The particle sizes and zeta potentials were measured using a Zetasizer Nano S «Malvern» (Worcestershire, UK) (4 mW He–Ne-laser, 633 nm, scattering angle 173°). The experiment was performed in a temperature-controlled cell at 25 °C. Autocorrelation functions of intensity fluctuations of light scattering were obtained using the correlation of the Correlator system K7032-09 «Malvern» (Worcestershire, UK). Experimental data were processed using «Zetasizer Software» (v. 8.02).

#### 2.2.3. Nanoparticle Tracking Analysis (NTA)

Determination of the hydrodynamic diameter of the synthesized polymers was carried out by NTA using a Nanosight LM10-HS device (Great Britain). Samples were diluted with MilliQ purified water to a concentration of $10^7$–$10^9$ particles/mL and kept in an ultrasonic bath for 30 s. The hydrodynamic diameter was determined by the Stokes–Einstein equation relating to the analysis of the trajectory of the Brownian motion of particles. Each sample was measured three times. The hydrodynamic diameter of the particles was also determined using the method of dynamic light scattering.

### 2.3. Drug Loading

Loading of model fluorophores-antibiotics and adjuvants (eugenol) into HPCD-PEI1.8 delivery systems was carried out by 2-h incubation at 50 °C (0.005 M HCl)—a five–fold mass excess of polymer over the drug.

### 2.4. FTIR Spectroscopy Studying of the Antibiotic's Actions on *E. coli* or CD206+ Macrophages Cells

*Escherichia coli* JM109 cells (overnight culture in liquid nutrient medium Luria–Bertani (pH 7.2), $10^8$ CFU) were washed twice with 0.01 M sterile PBS from the culture medium by centrifuging (Eppendorf centrifuge 5415C, 10 min, 12,000× *g*). Cell suspensions ($10^7$ CFU/mL) were incubated with antibiotic samples; then, after 1-2-12-24 h, the cell's samples were suspended and aliquots of 0.5 mL were taken. The cells are precipitated by centrifugation and separated from the supernatant, washed twice, and resuspended in 50 μL PBS to register the IR spectra. The supernatant is separated to determine the amounts of unabsorbed substances. ATR-FTIR spectra of cells samples suspension were recorded using a Bruker Tensor 27 spectrometer equipped with a liquid nitrogen-cooled MCT (mercury cadmium telluride) detector. Samples were placed in a thermostatic cell BioATR-II with a ZnSe ATR element (Bruker, Bremen, Germany). The FTIR spectrometer was purged with a constant flow of dry air (Jun-Air, Michigan, USA). FTIR spectra were acquired from 900 to 3000 $cm^{-1}$ with 1 $cm^{-1}$ spectral resolution. For each spectrum, 50–70 scans were accumulated at a 20 kHz scanning speed and averaged. Spectral data were processed using the Bruker software system Opus 8.2.28 (Bruker, Bremen, Germany), which includes linear blank subtraction, baseline correction, differentiation (second order, 9 smoothing points), min-max normalization, and atmosphere compensation. When necessary, 11-point Savitsky–Golay smoothing was used to remove noise. Peaks were identified by the standard Bruker picking-peak procedure. The concentration of Rif inside the cells was calculated from the material balance considering the unabsorbed Rif by UV-vis spectroscopy.

### 2.5. Macrophages Cell Lines

For the macrophage phagocytose assay, a human monocyte cell line THP-1 was used. Cells were obtained from the bank of cell lines at Lomonosov Moscow State University. THP-1 cells were cultured on T25 flasks in 5 mL RPMI-1640 (Gibco, Carlsbad, CA, USA), supplemented with GlutaMAX™ supplement (Gibco, Carlsbad, CA, USA) and buffered with 10 mM HEPES pH 7.4 containing 10% heat-inactivated FBS (Gibco, Carlsbad, CA, USA) and 1% antimycotic antibiotic (HyClone) at 37 °C and 5% $CO_2$. To derive macrophage-like cells, THP-1 cells were seeded on 6-well plates in 2 mL of RPMI-1640 (Gibco, Carlsbad, CA, USA), supplemented with GlutaMAX™ supplement (Gibco, Carlsbad, CA, USA) and buffered with 10 mM HEPES pH 7.4 containing 10% heat-inactivated FBS (Gibco, Carlsbad, CA, USA) and 1% antimycotic antibiotic (HyClone) with the addition of 100 nM phorbol 12- myristate 13-acetate (PMA, p8139, Sigma Aldrich, St. Louis, MI, USA) for 72 h. After 72 h, the medium containing PMA was replaced with RPMI-1640 (composition described above) without PMA and cells were cultured for another 96 h.

CD206-evaluation. To block nonspecific binding sites, cells were incubated with a 10% solution of normal goat serum in PBS with 1% bovine serum albumin BSA for 1 h at RT. Then, the samples were incubated with a solution of anti-CD206 antibodies (ab64693, Abcam, 1:100) or rabbit polyclonal control IgG (910801, Biolegend) as a control for 2 h at RT and subsequently with goat-anti-rabbit antibody conjugated with Alexa594 (A11037, Invitrogen, 1:1000). The nuclei were labeled with DAPI (Sigma-Aldrich, St. Louis, MO,

USA). Samples were analyzed with a Leica DM6000B fluorescent microscope equipped with a Leica DFC 360FX camera (Leica Microsystems GmbH, Wetzlar, Germany).

### 2.6. Confocal Laser Scanning Microscopy

*Escherichia coli* JM109 cells (overnight culture in liquid nutrient medium Luria–Bertani (pH 7.2), $10^8$ CFU) were centrifuged twice (Eppendorf centrifuge 5415C, 10 min, 12,000× *g*) and washed with 0.01 M PBS from the culture medium. Next, the cells were incubated for 60 min at 37 °C with 1 μg/mL of eosin-5-maleimide solution followed by twice-washing. Macrophages (CD206+ human monocyte cell line THP-1) placed in a 96-well fluorescent plate (Costar) were incubated for 1 h with eosin-labeled bacteria followed by washing (10 min, 4000× *g*). Samples (Dox in free form and with FITC-labeled delivery systems) were added to macrophages with absorbed *E. coli*. The cells were centrifuged twice with PBS washing (10 min, 4000× *g*). The cell centrifuge was suspended in 200 μL of PBS, followed by the addition 100 μL of a 5% agarose solution at 45 °C to solidify the cell suspension in the wells of a fluorescent plate. Fluorescence images were obtained by the confocal laser scanning microscope (CLSM) Olympus FluoView FV1000 equipped with both a spectral version scan unit with emission detectors and a transmitted light detector. CLSM is based on the motorized inverted microscope Olympus IX81. Emission fluorescence spectra of FITC (drug delivery system labelled), eosin (*E. coli* labelled), and Dox was obtained by CLSM. The excitation wavelength 488 nm (multiline Argon laser) and dry objective lens Olympus UPLSAPO 40X NA 0.90 were used for the measurements. Laser power, sampling speed, and averaging were the same for all image acquisitions. The scan area was $80 \times 80\ \mu m^2$. FITC, Eosin, and Dox fluorescence was collected using the emission windows set at 505–540, 540–575 nm, and 575–675, respectively, at 488 nm excitation. The signals were adjusted to the linear range of the detectors. Olympus FV10 ASW 1.7 software was used for acquisition of the images. FITC fluorescence is shown in green, Dox is red, Eosin is magenta, and the image on the light is gray.

### 2.7. Dox, FITC-Labelled Ligand, and Eosin-Labelled E. coli Determination Macrophage Uptake

Quantitative analysis of Dox, FITC-labelled ligand (as in paper [50]), and eosin-labelled *E. coli* (Section 2.6) content in CD206+ macrophages was performed using fluorescence spectroscopy. $\lambda_{exci}$ (Dox or FITC) = 490 nm. $\lambda_{exci}$ (eosin) = 515 nm. $\lambda_{emi}$ (Dox) = 595 nm, $\lambda_{emi}$ (eosin) = 560 nm, $\lambda_{emi}$ (FITC) = 520 nm. Registration of fluorescence spectra was carried out using a SpectraMax M5 device (Pennsylvania, USA) in the Costar black/clear bottom tablet (96 wells). T = 25 °C. The concentration of Dox, FITC, and eosin inside the cells was calculated from the material balance considering the unabsorbed fluorophore's concentration determined by fluorescence intensity. Intracellular concentrations of fluorophores were determined after destruction of macrophage cells by 10-min incubation with 1% Triton X-100 solutions.

### 2.8. Antibacterial Activity of Rif

The strain used in this study was *Escherichia coli* JM109 (J.Messing, USA). The culture was cultivated for 18–20 h at 37 °C to CFU $\approx 1.5 \times 10^8$–$2 \times 10^8$ (colony-forming unit, determined by A600) in the liquid nutrient medium Luria–Bertani (pH 7.2) without stirring. The experiments in liquid media were conducted by adding 50 μL of the samples in 5000 μL of cell culture. The specimens were incubated at 37 °C for seven days. At the specific time, 100 μL of each sample was taken, diluted with distilled water, and the absorbance was measured at 600 nm. For quantitative analysis, the dependences of CFU (cell viability) on the concentration of Rif, 50 μL of each sample was diluted $10^5$–$10^9$ times and seeded on the Petri dish. Dishes were placed in the incubator at 37 °C for 24 h. Then, the number of the colonies (CFU) was counted.

## 2.9. Statistical Analysis

A statistical analysis of the obtained data was carried out using the Student's *t*-test Origin 2022 software (OriginLab Corporation, Northampton, MA, USA). Values are presented as the mean ± SD of three experiments (three replicates).

# 3. Results and Discussion

## 3.1. FTIR Spectroscopy of E. coli—Drug Interaction's Tracking

FTIR spectroscopy can be effectively used to monitor the molecular details of the interactions of medicinal preparations with cells. In the cell, it is possible to distinguish the main structural units that contribute to the absorption of IR radiation (Figure 1): cell membrane lipids (2800–3000 $cm^{-1}$), proteins, especially transmembrane (1500–1700 $cm^{-1}$), DNA phosphate groups (1240 $cm^{-1}$), and carbohydrates, including lipopolysaccharides (900–1100 $cm^{-1}$). The main fluctuations of bonds in the structural units of E. coli cells were: 2960–2850 $cm^{-1}$ CH, $CH_2$, $CH_3$ in fatty acids, 1655–1637 $cm^{-1}$ amide I bands (α-helical and β-pleated sheet structures), 1548 $cm^{-1}$ amide II band, 1515 $cm^{-1}$ aromatic band, 1465–1470 C–H deformation, 1310–1240 $cm^{-1}$ amide III band components of proteins, 1250–1220 and 1084–1088 $cm^{-1}$ P=O stretching of $PO_2^-$ – phosphodiesters, and 1100–900 $cm^{-1}$ C–O–C, C–O of saccharide ring vibrations [23]. The IR spectrum of lipids and phospholipids has the following characteristic peaks of functional groups: two bands of symmetric and asymmetric vibrations of hydrocarbon bonds, vibrations of the carbonyl group C=O, and vibrations of the phosphate (Figure 1). The position of the bands and their shape are sensitive to binding of the bilayer with ligands or drug molecules, hydrogen bond formation, aggregation, and oxidation, etc. [64].

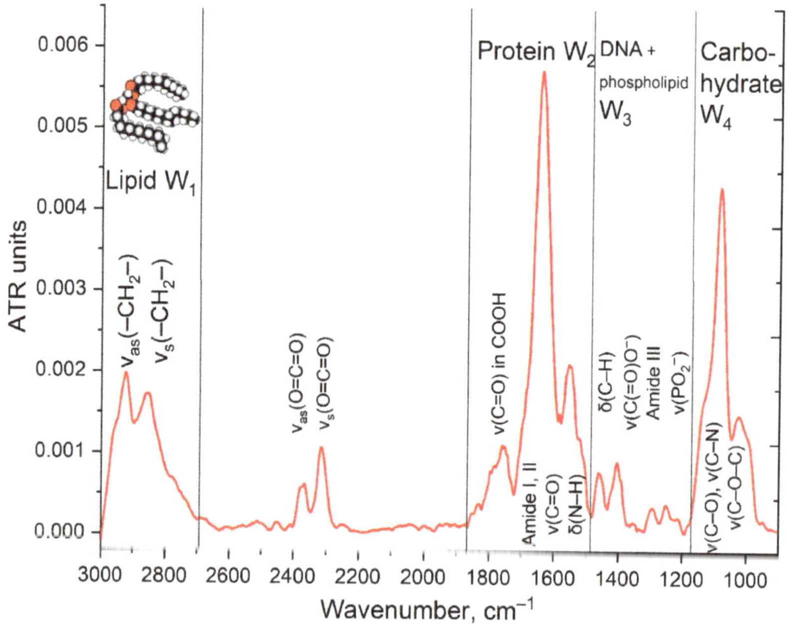


**Figure 1.** FTIR spectra of *E. coli* cells suspension in water. T = 22 °C.

To enhance the antibiotics efficiency, we used polymer nanoparticles HPCD-PEII1.8-triMan (polyethyleneimines grafted with cyclodextrins and with a carbohydrate labels on CD206 macrophage receptors (Table 1), as well as an adjuvant (on the example of eugenol), which inhibits the pumping of drugs from cells and increases bioavailability [6,50].

**Table 1.** Characteristics of drug delivery systems and their affinity to mannose receptor.

| Carrier * | Label (n) ** | Molecular Weight, kDa | Hydrodynamic Size ***, nm | Zeta-Potential ****, mV | Polydispersity Index |
|---|---|---|---|---|---|
| HPCD-PEI1.8-FITC * (3:1: 0.5:n) ** | Man (12) | 9 ± 3 | 176 ± 100 | −4 ± 2 | 0.4 |
| | triMan (3) | | 120 ± 50 | −6.5 ± 1.5 | 0.45 |
| | Gal (12) | | 170 ± 60 | −7 ± 3 | 0.4 |

* FITC-labeled ligands were used only for experiments with macrophages. ** n is the number of carbohydrate labels. *** by NTA. **** by DLS.

Rif and EG are only poorly soluble in water, so they need to be included in the delivery system, and the simplest is methyl-cyclodextrin (MCD). Further, the authors use cyclodextrin to prove the effectiveness of polymeric conjugates grafted by CD vs. simple cyclodextrin. Figure 2 shows the difference FTIR spectra (the spectrum is subtracted at zero time) of *E. coli* cell suspensions incubated with free Rif, Rif as part of a molecular container, EG in the form of an inclusion complex with β-cyclodextrin, and a combined formulation of antibiotic and adjuvant loaded into the delivery system. The aim of the experiment is to study the influence of the concentration of substances and the incubation time with cells on the changes in IR spectra of the cells, in other words, how the interaction of drugs and polymers with bacteria and macrophages is reflected on the spectra. The most pronounced changes are observed in the absorption bands of amides 1 and 2 (1600–1700 and 1500–1600 $cm^{-1}$), oscillations of $CH_2$ groups (2800–3000 $cm^{-1}$), as well as in the region of 1240 $cm^{-1}$ ($PO_2^-$ phospholipids and DNA) and 1000–1100 $cm^{-1}$ (C-O-C carbohydrates).

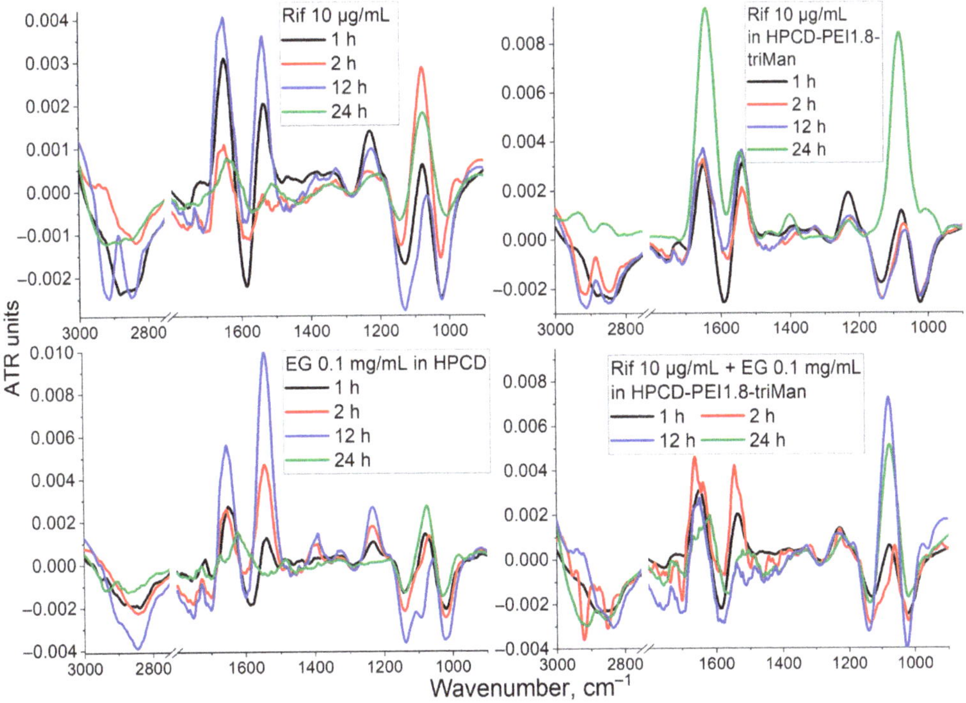


**Figure 2.** Difference FTIR spectra of *E. coli* incubated with rifampicin (Rif) or/and eugenol (EG) drug formulations. The spectra at the zero moment of time are subtracted. T = 22 °C.

Changes in the W1 region (Figures 1 and 2) correspond to a change in the structural organization of the membrane and the accumulation of an antibiotic or EG inside the cells. For free Rif (Figure 2) in the first hour, there is a significant ($p$-value = 0.014) decrease in the intensity of ATR in W1, which indicates the incorporation of hydrophobic Rif molecules into the bacterial membrane (disordering of lipids). After 2 h of incubation, the antibiotic has already begun to accumulate inside the cells (the difference intensity in the IR spectrum decreases dramatically in the W1 region), which correlates with the data that the Rif penetrates into cells after 2–3 h (Figure 3). After 24 h, the antibiotic was eliminated by more than 70% cells (Figure 3), most probably because of efflux (quantitative determination of efflux was carried out by us earlier) [6]. The Rif concentration and ability to interact with transmembrane proteins correlate with the intensity of amide peaks 1 and 2 (Figure 2).

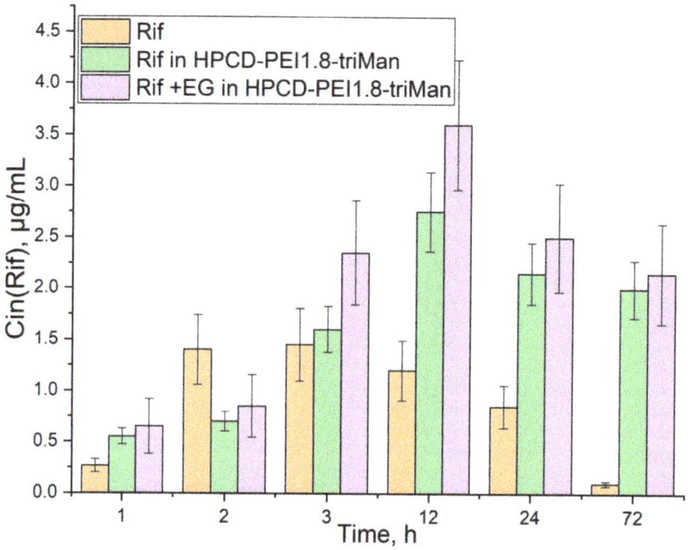


**Figure 3.** Intracellular (*E. coli*) concentrations of Rif pre-incubated with cells in the form of various formulations: free form, in the delivery system, and enhanced with the adjuvant (EG).

The inclusion of antibiotics in the delivery system to CD206+ macrophages (without EG) leads to significant ($p < 0.003$) effects on the accumulation of drugs inside cells (Figure 3). Accelerated drug absorption compared to free form is observed: (1) owing to the adsorption of polymer particles on the membrane surface (0.004–0.009 ATR vs. 0.001–0.004 in W2) and (2) owing to the occurrence of local defects in the membrane and increased penetration of the drug into the bacteria. In addition, the prolonged action of Rif in the delivery system is achieved, which can be observed by the changed in the intensity of amide 1 (Figure 2, top row). After 12–24 h, free Rif is characterized by a low intracellular concentration (Figure 3), and the drug in polymer particles is still working for 2–4 days [65,66]. The FTIR spectra of *E. coli* incubated with drug delivery system HPCD-PEI1.8-triMan itself are presented in Figure S1; the observed increase in intensity in the W2 region indicates the polymer carrier interacts with transmembrane proteins.

The interaction of EG–MCD (efflux inhibitor and enhancing membrane permeability agent [6]) with bacterial cells (Figure 2, bottom row) leads to: (i) inhibition of efflux pumps (as can be judged from increase in the intensity of amides 1 and 2) and (ii) the creation of defects (earlier was shown using CLSM [6]), which is reflected in a decrease W1 FTIR intensity and increase W3, corresponding to DNA and phospholipids). This explains the synergy of EG with antibiotics: we previously showed for levofloxacin and moxifloxacin enhancing antibacterial activity and we found enhanced absorption by CLSM in the sample with EG [6,20,50,63]. Similar effects are observed here for rifampicin by FTIR. The greatest effect is achieved for the combined antibiotic and adjuvant system in a polymer carrier (Figure 2, bottom row): strong amplification of W2 and W3 peaks, which correlates with blocking of efflux and high intracellular Rif concentration (Figure 3).

The changes discussed above in the FTIR spectra of *E. coli* when interacting with antibacterial agents reflect the accumulation of Rif inside cells (Figure 3). The concentration of free Rif does not exceed 1.5 μg/mL (calculated from the material balance considering the unabsorbed Rif by UV-VIS spectroscopy) and drops after 2–4 h of incubation. However, in the complex polymeric formulation, the penetration of the antibiotic is much more effective: a concentration of >2.5–3.5 μg/mL is achieved and, moreover, it is maintained at >2 μg/mL for 72 h.

### 3.2. FTIR Spectroscopy of E. coli in CD206+ Macrophages—Drug Interaction's Monitoring

The CD206 mannose receptor is of greatest interest, which is involved in the recognition of pathogens stemming from the interaction of protein-binding domains with oligosaccharide patterns of microorganisms (*Candida albicans, Pneumocystis carinii, Leishmania donovani, Mycobacterium tuberculosis, Klebsiella pneumoniae,* etc.) [67,68]. The CD206 receptor mainly allows for targeting activated macrophages, in which resistant and dormant infections can accumulate. Selectivity toward micro-organisms is achieved because of the specificity of CD206 to mannose, fucose, and N-acetylglucosamine residues, which often cover the surface of pathogen cells, unlike mammals [31,41,49,56,69].

As shown above, the HPCD-PEI1.8-triMan molecular container and eugenol adjuvant enhance Rif penetration into bacterial cells and cause prolonged action according to FTIR spectroscopy data. Pathogens are least accessible when they localized in macrophages, so it is necessary to deliver antibacterial agents to macrophages, for example through the CD206 receptor.

Figure 4 shows the FTIR spectra of CD206+ macrophages with bacteria absorbed by them. We studied the interaction of macrophages with polymer carriers with three carbohydrate vectors of different affinity to CD206 (galactose—with low affinity to CD206, mannose with medium affinity and high affinity trimannoside vector), as well as the use of the EG adjuvant using FTIR spectroscopy, orthe effect of phagocytosis on the spectra. As shown earlier by flow cytometry [6,50] and confirmed here with FTIR spectroscopy (Figure 4), macrophages phagocytize polymer particles mainly with a high-affinity vector (triMan). Changes in the membrane of macrophages and E. coli are reflected in the region of 3000–2850 $cm^{-1}$: the highest intensity means increased phagocytic activity and, consequently, greater accessibility for bacterial cells, which is further confirmed by an increase in the intensity of the peak of 1150–1000 $cm^{-1}$ corresponding to the number of polymers adsorbed on *E. coli* and absorbed by macrophages. The CD206+ dependent binding of drug delivery systems to macrophages is confirmed by quenching peaks of amides 1 and 2, which is typical only for mannose-labeled polymers. Eugenol additionally enhances the accumulation of only high-affinity ligands (bottom row, Figure 4—amide region 1 and 2) and thereby increases the selectivity of the developed HPCD-PEI1.8-triMan carriers.

### 3.3. CLSM of E. coli in CD206+ Macrophages—Drug Interaction's Visualization

To clarify the action mechanisms of polymeric carriers and adjuvant, CLSM and fluorescent studies of the drugs interaction with bacteria were carried out. We made a model system of macrophages with absorbed *E. coli*, which are colored with eosin, to study phagocytosis by macrophages of FITC-labeled HPCD-PEI1.8-triMan loaded with the fluorophore—antibiotic doxorubicin (Dox).

We studied three groups of samples:

(1) Control Dox to study the penetration and accumulation of free drug in macrophages and inside bacterial cells;

(2) Dox in a polymeric ligand (with different CD206-affinity labels: non-specific galactose, medium-affine mannose and high-affine triMan) to study macrophage phagocytosis

activity and the effect of polymer on the adsorption efficiency on the bacterial cells and penetration of Dox;

(3) Dox in a polymeric ligand enhanced with EG as an agent enhancing the membrane permeability and efflux inhibitor.

The assignment of fluorescence signals (Dox, eosin, and FITC) is based on the fluorescence spectra of substances in the systems under consideration (Figure S2, Method section). Confocal images of macrophages with insider bacteria are shown in Figure 5 and S3. On confocal images, large macrophages can be observed, inside of which *E. coli* are highlighted in pink, in which Dox accumulates asred dots.

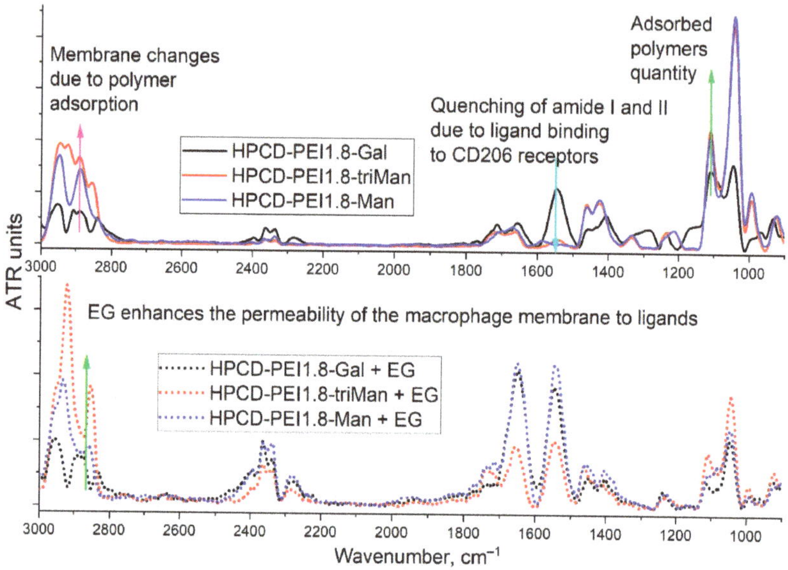


**Figure 4.** FTIR spectra of CD206+ Macrophages with absorbed E. coli incubated with drug delivery systems with different carbohydrate vectors and eugenol-enhanced formulations.

Free Dox accumulates weakly in macrophages and inside *E. coli* (Figure 5a). Dox is released from the bacteria by pumping proteins in a process of efflux, and macrophages, in principle, poorly absorb small drug molecules.

The effect of adjuvant EG (Figure 5b) *on the accumulation of Dox.* An increase in the degree of Dox accumulation in bacterial cells in macrophages is observed. EG acts in two directions: creates defects in the membrane and inhibits pump proteins, as we showed in the last article [6].

The effect of drug delivery systems on the accumulation of Dox (Figure 5c–e). CD206-positive cells effectively phagocytosed predominantly high-affinity polymeric conjugate with trimannoside HPCD-PEI1.8-triMan, but not conjugate with linear galactose, which follows from the intensity of macrophage-associated fluorescence in the FITC channel (green). The label of linear mannose on the conjugate is medium effective. Inside the macrophages, colored dots are visible (in all channels), corresponding to the bacteria on which the polymer is adsorbed. Owing to the high penetration of HPCD-PEI1.8-triMan into macrophages, the accumulation of Dox in *E. coli* is very high, relative to control samples (Dox). Earlier, our cytometry assay determined that 80% of macrophage-like cells were FITC-positive after adding HPCD-PEI1.8-triMan, 60% were FITC-positive after adding HPCD-PEI1.8-Man, and 15% were FITC-positive after adding HPCD-PEI1.8-Gal [50]. Therefore, the data on CLSM and FTIR correlate with the flow cytometry data.

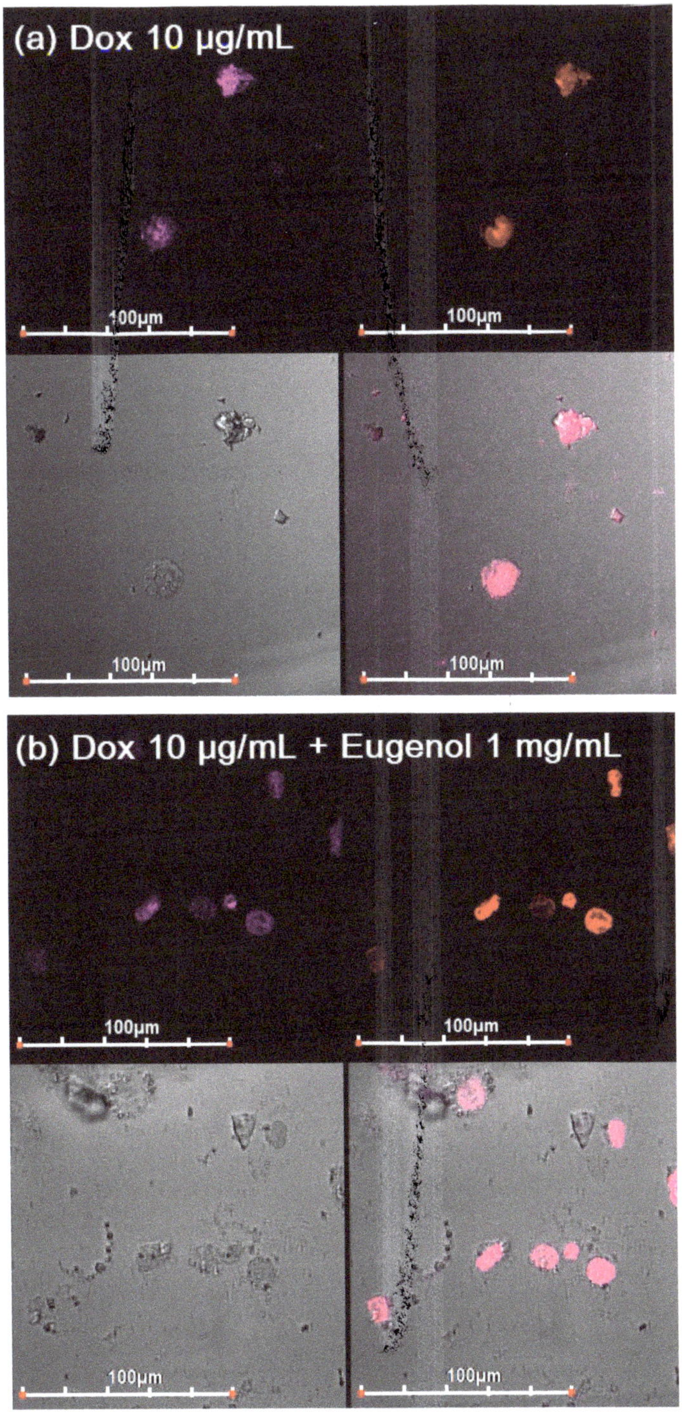


**Figure 5.** *Cont.*

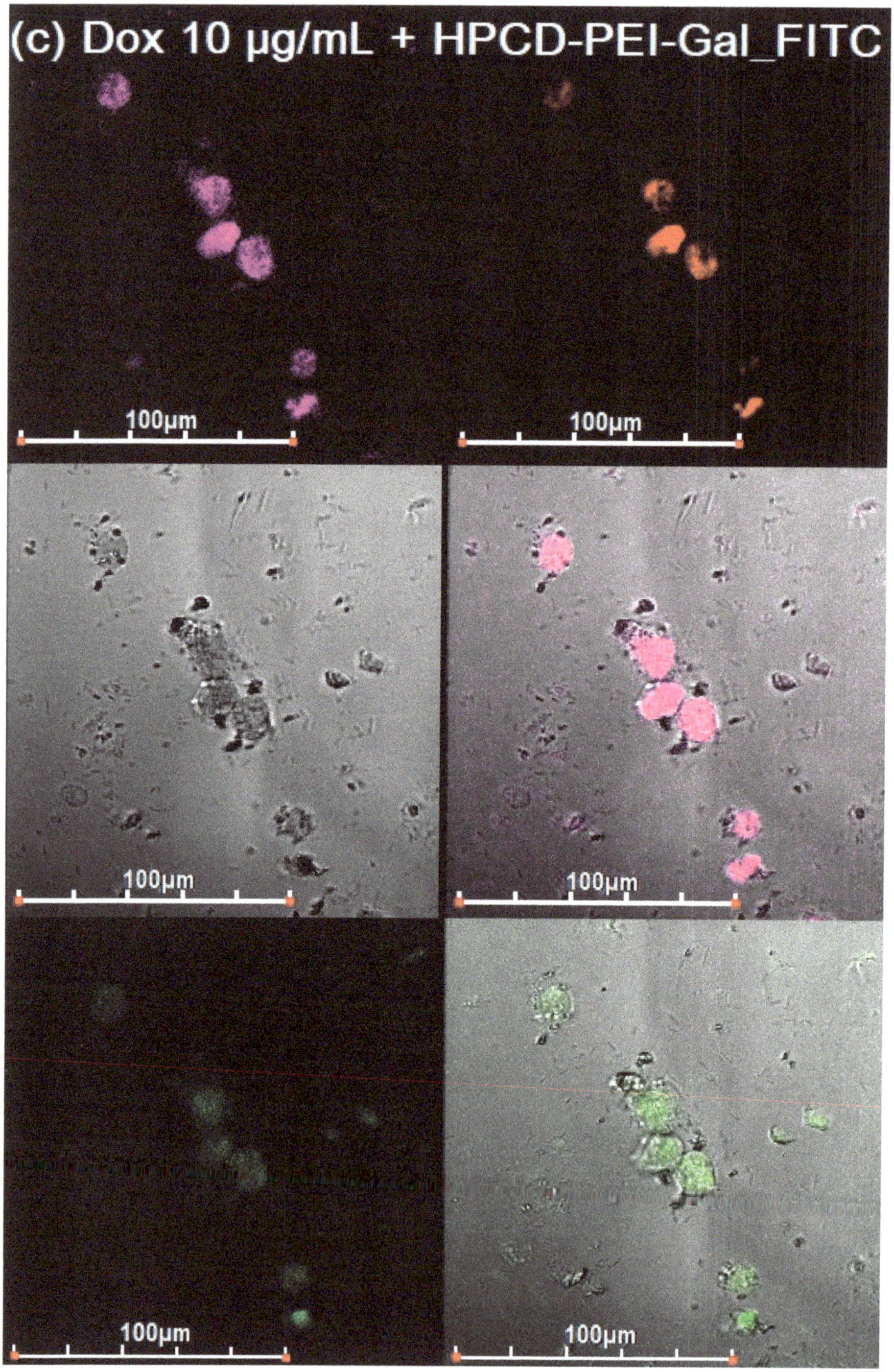


**Figure 5.** *Cont.*

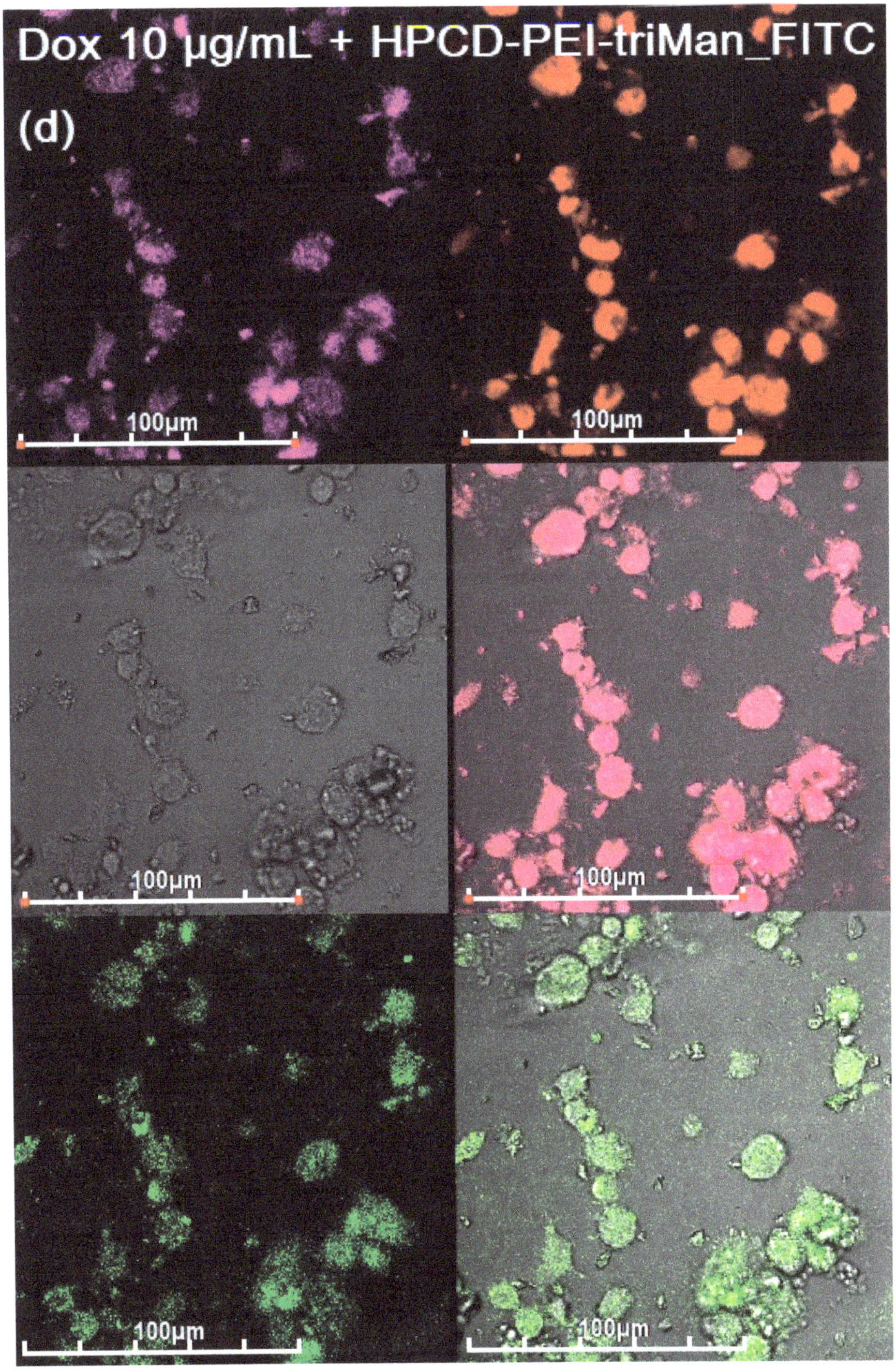


**Figure 5.** *Cont.*

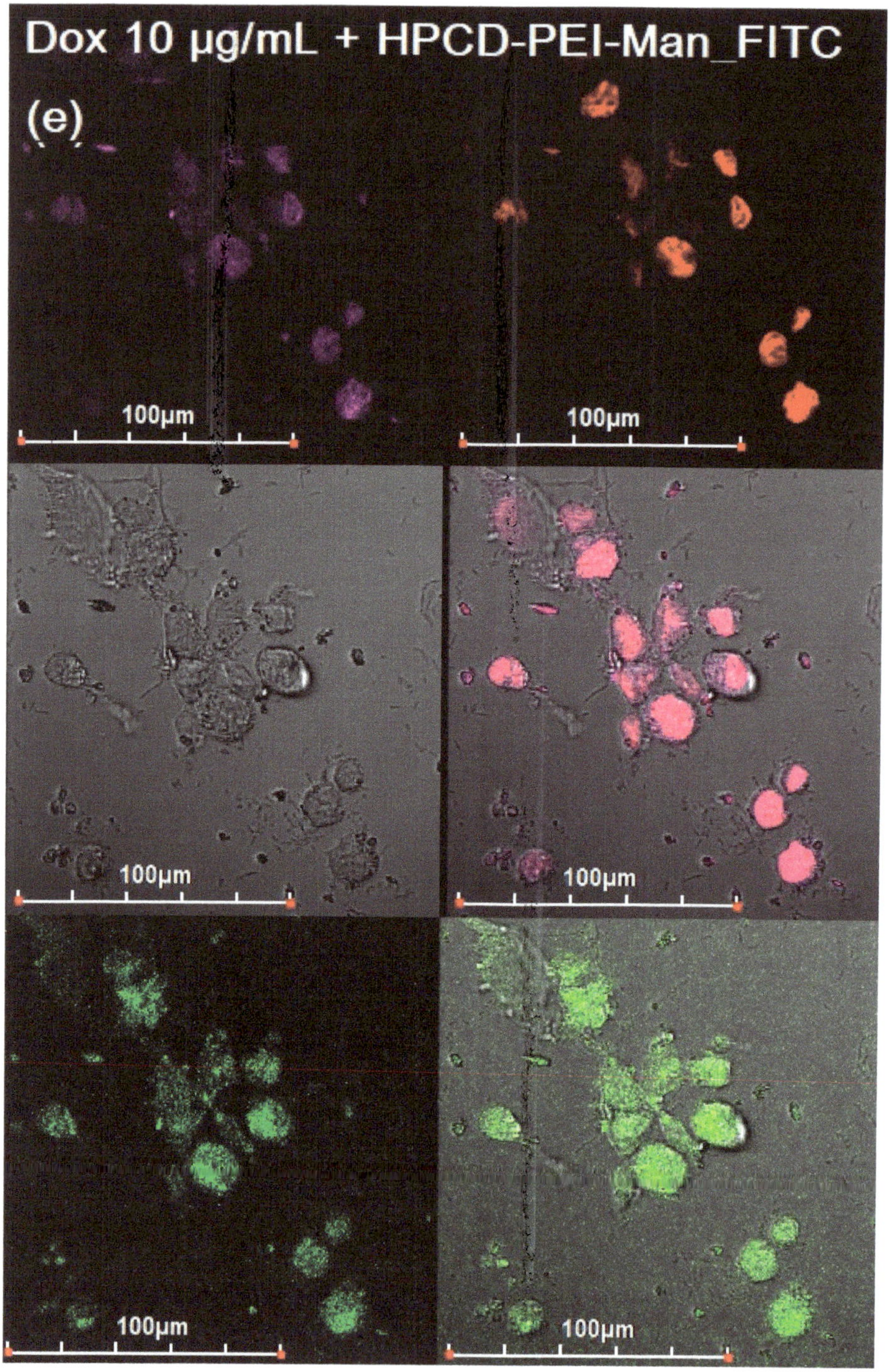


**Figure 5.** *Cont.*

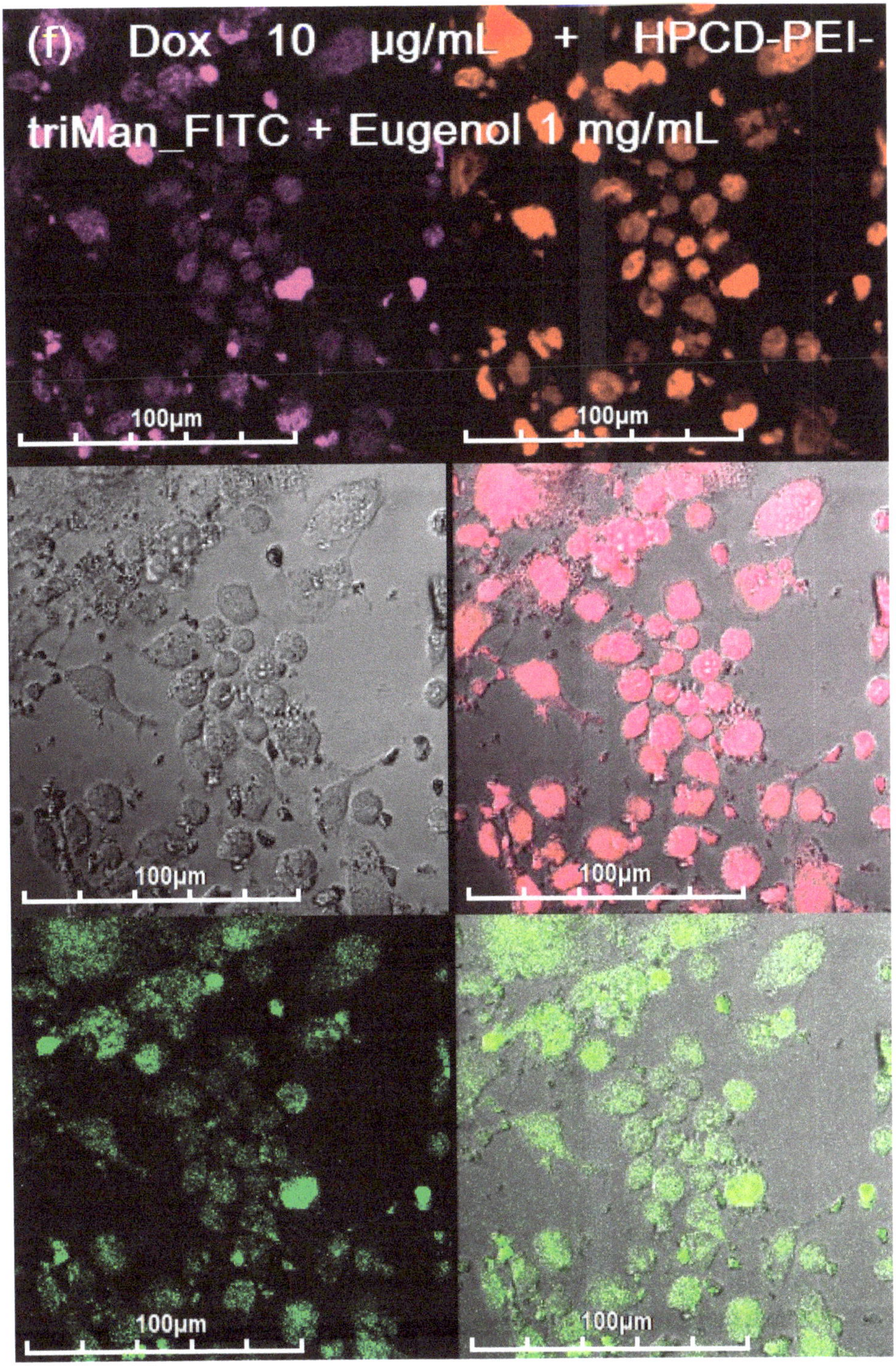


**Figure 5.** Confocal laser scanning images of CD206+ macrophages with absorbed eosin-labelled *E. coli*.

Incubation 2 h with Dox 10 µg/mL and FITC-labeled HPCD-PEI1.8. Dox 10 µg/mL: (**a**) free, (**b**) with 1 mg/mL EG, (**c**) in FITC-labelled HPCD-PEI1.8-Gal, (**d**) in FITC-labelled HPCD-PEI1.8-triMan, (**e**) in FITC-labelled HPCD-PEI1.8-Man, (**f**) in FITC-labelled HPCD-PEI1.8-triMan with 1 mg/mL EG. The scale segment is 100 µm (division value is 20 µm); 4–6 channels are shown: red, Dox; green, FITC; magenta, eosin; gray, transmission light mode; and overlay. $\lambda_{em}$ = 488 nm (multiline Argon laser).

The synergy effect antibiotic and adjuvant, loaded into polymeric nanoparticles (Figure 5f and S3). Dox in combination with eugenol, which enhances penetration through cell membranes and inhibits drug pumping, accumulates 10 times more efficiently than a simple substance in the composition of high-affinity conjugates to CD206+ macrophages. Thus, the complex formulation antibiotic and adjuvant in a drug delivery system is a perfect approach to accumulate the drug in the target macrophages with pathogenic bacteria.

In summary, by using CLSM and FTIR, we were able to distinguish bacteria inside macrophages, and showed how ligands penetrate into macrophages and thereby increase the accumulation of drugs inside bacteria.

Thus, confocal microscopy confirms the data of FTIR spectroscopy in the terms of enhancing the permeability of the bacterial membrane to the drug caused by polymers and adjuvants (efflux inhibitors) and phagocytosis by macrophages. Therefore, spectral changes are confirmed visually and quantitatively (Section 3.4. Table 2).

**Table 2.** The amounts of Dox and FITC-labeled carrier absorbed by macrophages with E. coli inside. Fluorescence detection (Method section). T = 22 °C.

| Sample | | | Dox Absorbed, % | FITC-Labeled Carrier Absorbed, % |
|---|---|---|---|---|
| Dox 10 µg/mL | - | - | 23 ± 2 | - |
| | | Eugenol 1 mg/mL | 30 ± 3 | |
| | HPCD-PEI1.8-Gal | | 27 ± 4 | 13 ± 4 |
| | HPCD-PEI1.8-triMan | | 65 ± 4 | 73 ± 7 |
| | HPCD-PEI1.8-Man | | 49 ± 3 | 53 ± 3 |
| | HPCD-PEI1.8-Gal | - | 27 ± 5 | 12 ± 1 |
| | HPCD-PEI1.8-triMan | | 77 ± 3 | 63 ± 4 |
| | HPCD-PEI1.8-Man | | 53 ± 2 | 57 ± 2 |

### 3.4. *Quantitative Data on the Penetration of Drugs into Macrophages with E. coli*

Table 2 presents quantitative data on the absorption of Dox and FITC-labeled conjugates (determined based on the material balance of extracellular and intracellular concentrations in macrophages after lysis with 1% Triton X-100, fluorescent detection). The data on ligand uptake correlate with those previously obtained by flow cytometry: the carrier with the trimannoside vector is absorbed by cells by more than 60–70%, and galactose-labeled by only 10–15%. Thus, owing to polymer ligand and adjuvants (eugenol or its analogues), it is possible to increase the accumulation of Dox inside macrophages by more than three times. Taking into account the previously obtained values for isolated *E. coli* [6], the accumulation of Dox directly in the bacteria increased by more than 10 times (it implies the total effect of enhanced penetration into macrophages x 3–4 and then in *E. coli* inside macrophages × 3).

### 3.5. *Rif Antibacterial Activity on E. coli*

Using FTIR spectroscopy, changes in cells during incubation with drugs have been demonstrated—therefore, it is important to show how these data correlate with antibacterial activity. The antibacterial activity of Rif in a free form and in the composition of molecular containers and enhanced with eugenol demonstrates correlations between the observed effects using spectral and optical methods and what is actually observed in a microbiological experiment. Figure 6 shows the curves of the survival of *E. coli* bacterial cells on the incubation time with various forms of Rif in comparison with the control (without the addition of an antibiotic). Apparently, polymers accelerate the penetration of antibiotics into cells because of the adsorption of polysaccharides on the cell wall surface. According to the data on CFU and the turbidity of the cell suspension (A600), the HPCD-PEI1.8-triMan polymer itself does not affect cell growth, and the Rif efficiency practically does not increase when incorporated into cyclodextrin (MCD); however, when loaded into the polymer system, the efficiency increases significantly (0.5–1 CFU order). Eugenol in the form of a complex with MCD demonstrates antibacterial activity; however, it is not bright in itself. At the same time, eugenol acts as a synergist as an adjuvant to the antibiotic Rif, so cell growth is quickly suspended, and the number of viable cells falls. Previously, the synergism of antibiotics (moxifloxacin and levofloxacin) with eugenol, menthol, and apiol was demonstrated by us [20,63,70]. Thus, the complex formulation of antibiotic and booster adjuvant in the delivery system with active targeting function is a promising combination for the possible strengthening of existing therapeics.

Thus, the correlations of the antibacterial activity of therapeutic agents with the observed effects in the FTIR spectra and visually in a confocal microscope are demonstrated. Optical methods are applicable for the diagnosis and optimization of the treatment of bacterial infections.

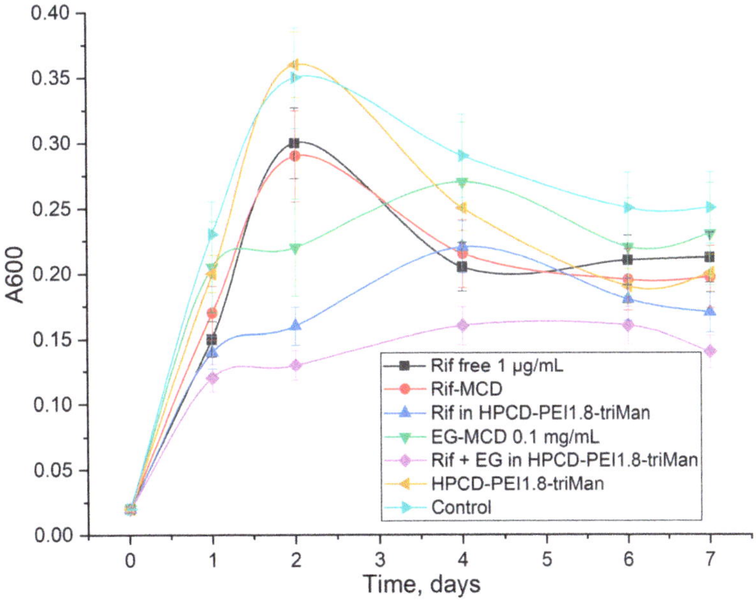


(a)

**Figure 6.** *Cont.*

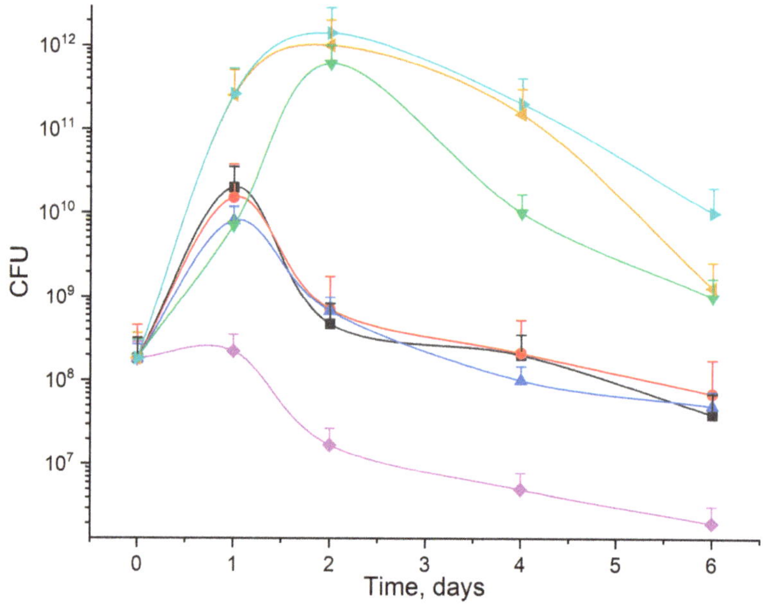


(**b**)

**Figure 6.** Antibacterial activity of Rif (1 µg/mL) in free form and loaded into HPCD-PEI1.8-triMan, enhanced with eugenol (0.1 mg/mL): (**a**) absorbance of samples (in terms of the diluted samples (absorption was determined in aliquots diluted 4–10 times)) correlated with CFU, and (**b**) CFU dependences determined by quantitative seeding on petri dishes. Rif: polymer mass ratio = 1:10. pH 7.4 (0.01 M PBS), 37 °C.

## 4. Conclusions

Visualization of the action of therapeutic agents at the molecular level seems to be a powerful way to find out the points of increasing the effectiveness of antibacterial drugs, and in the future, cytostatic ones. The methods of FTIR spectroscopy and confocal microscopy are potentially applicable for the diagnosis of latent infections localized in macrophages, such as pneumonia, tuberculosis, and mycoplasma. Using FTIR spectroscopy, it was shown that the accumulation of rifampicin in model *E. coli* cells increases with the use of polymeric molecular containers and adjuvants that inhibit efflux and increase membrane permeability. The penetration of the model fluorophore antibiotic doxorubicin into CD206+ macrophages with *E. coli* localized in them was visualized using CLSM. The efficiency of phagocytosis of polymer drug carriers by macrophages depending on the carbohydrate label (galactose, mannose, trimannoside) was compared. The developed delivery systems increase the effectiveness of the therapeutic agent (Dox or Rif) by more than 10 times (it implies the total effect of enhanced penetration into macrophages ×3–4 and then inside macrophages in *E. coli* ×3). Thus, we presented the potential of practical application of optical and spectral methods (FTIR + CLSM) in aspects of drug study and possible diagnosis of diseases.

**Supplementary Materials:** The following supporting information can be downloaded at: https://www.mdpi.com/article/10.3390/diagnostics13040698/s1, Figure S1. FTIR spectra of *E. coli* incubated with drug delivery system HPCD-PEI1.8-triMan. T = 22 °C. Figure S2. Emission fluorescence spectra of FITC (drug delivery system labelled), eosin (*E. coli* labelled), and Dox obtained by CLSM. $\lambda_{em}$ = 488 nm (multiline Argon laser). **Figure S3**. Confocal laser scanning images of CD206+ macrophages with absorbed eosin-labelled *E. coli*. Incubation with Dox and FITC-labeled

HPCD-PEI1.8-triMan. The scale segment is 100 μm (division value is 20 μm); 4–6 channels are shown: red, Dox; green, FITC; magenta, eosin; gray, transmission light mode; and overlay $\lambda_{em}$ = 488 nm (multiline Argon laser).

**Author Contributions:** Conceptualization, I.D.Z. and E.V.K.; methodology, I.D.Z., A.A.E. and E.V.K.; software, M.A.V., N.G.B., U.D.D. and O.A.G.; formal analysis, I.D.Z. and A.A.E.; investigation, I.D.Z., A.A.E., N.G.B., M.A.V., U.D.D. and O.A.G.; data curation, I.D.Z.; writing—original draft preparation, I.D.Z.; writing—review and editing, E.V.K.; project administration, E.V.K.; funding acquisition, E.V.K. All authors have read and agreed to the published version of the manuscript.

**Funding:** This research was funded by the Russian Science Foundation, grant number 22-24-00604.

**Institutional Review Board Statement:** All procedures with the involvement of animals complied with the ethical standards approved by the legal acts of the Russian Federation, the principles of the Basel Declaration, and the recommendations of the Bioethics Committee at Lomonosov Moscow State University.

**Informed Consent Statement:** Not applicable.

**Data Availability Statement:** The data presented in this study are available in the main text and Supplementary Materials.

**Acknowledgments:** The work was performed using equipment (FTIR spectrometer Bruker Tensor 27, Scanning probe microscope NT-MDT and Jasco J-815 CD Spectrometer) of the program for the development of Moscow State University.

**Conflicts of Interest:** The authors declare no conflict of interest.

## Abbreviations

| | |
|---|---|
| CLSM | confocal laser scanning microscopy |
| CFU | colony-forming unit |
| Chit | chitosan |
| $C_{in}$ | inside concentration |
| Dox | doxorubicin |
| HPCD | 2-hydroxypropyl-β-cyclodextrin |
| FTIR | Fourier transform infrared (spectroscopy) |
| MCD | methyl-β-cyclodextrin |
| MM | molar mass |
| NTA | nanoparticle tracking analysis |
| Rif | rifampicin |
| triMan | trimannoside residue |
| EG | eugenol; |
| PEI | polyethyleneimine; |

## References

1. Esposito, S.; Bianchini, S.; Argentiero, A.; Neglia, C.; Principi, N. How Does One Choose the Appropriate Pharmacotherapy for Children with Lower Respiratory Tract Infections? *Expert Opin. Pharmacother.* **2020**, *21*, 1739–1747. [CrossRef]
2. Li, J.; Song, X.; Yang, T.; Chen, Y.; Gong, Y.; Yin, X.; Lu, Z. Systematic Review of Antibiotic Prescription Associated with Upper Respiratory Tract Infections in China. *Medicine* **2016**, *95*, 1–7. [CrossRef]
3. Levy, S.B.; Bonnie, M. Antibacterial Resistance Worldwide: Causes, Challenges and Responses. *Nat. Med.* **2004**, *10*, S122–S129. [CrossRef]
4. Garske, L.A.; Kidd, T.J.; Gan, R.; Bunting, J.P.; Franks, C.A.; Coulter, C.; Masel, P.J.; Bell, S.C. Rifampicin and Sodium Fusidate Reduces the Frequency of Methicillin-Resistant Staphylococcus Aureus (MRSA) Isolation in Adults with Cystic Fibrosis and Chronic MRSA Infection. *J. Hosp. Infect.* **2004**, *56*, 208–214. [CrossRef]
5. Malhotra-Kumar, S.; Lammens, C.; Coenen, S.; Van Herck, K.; Goossens, H. Effect of Azithromycin and Clarithromycin Therapy on Pharyngeal Carriage of Macrolide-Resistant Streptococci in Healthy Volunteers: A Randomised, Double-Blind, Placebo-Controlled Study. *Lancet* **2007**, *369*, 482–490. [CrossRef]
6. Zlotnikov, I.D.; Ezhov, A.A.; Petrov, R.A.; Vigovskiy, M.A.; Grigorieva, O.A.; Belogurova, N.G.; Kudryashova, E.V. Mannosylated Polymeric Ligands for Targeted Delivery of Antibacterials and Their Adjuvants to Macrophages for the Enhancement of the Drug Efficiency. *Pharmaceuticals* **2022**, *15*, 1172. [CrossRef]

7. Jiso, A.; Khemawoot, P.; Techapichetvanich, P.; Soopairin, S.; Phoemsap, K.; Damrongsakul, P.; Wongwiwatthananukit, S.; Vivithanaporn, P. Drug-Herb Interactions among Thai Herbs and Anticancer Drugs: A Scoping Review. *Pharmaceuticals* **2022**, *15*, 146. [CrossRef]
8. Neyfakh, A.A.; Bidnenko, V.E.; Chen, L.B. Efflux-Mediated Multidrug Resistance in Bacillus Subtilis: Similarities and Dissimilarities with the Mammalian System. *Proc. Natl. Acad. Sci. USA* **1991**, *88*, 4781–4785. [CrossRef]
9. Polyakova, A.; Bakradze, M.; Tatochenko, B.; Chashchina, I. Bacterial Infections Of The Upper Respiratory Tract —How To Treat? *Medical Council* **2018**, *17*, 94–102. [CrossRef]
10. Dalhoff, A. Global Fluoroquinolone Resistance Epidemiology and Implictions for Clinical Use. *Interdiscip. Perspect. Infect. Dis.* **2012**, *2012*, 976273. [CrossRef]
11. Redgrave, L.S.; Sutton, S.B.; Webber, M.A.; Piddock, L.J.V. Fluoroquinolone Resistance: Mechanisms, Impact on Bacteria, and Role in Evolutionary Success. *Trends Microbiol.* **2014**, *22*, 438–445. [CrossRef]
12. Peyron, P.; Vaubourgeix, J.; Poquet, Y.; Levillain, F.; Botanch, C.; Bardou, F.; Daffé, M.; Emile, J.F.; Marchou, B.; Cardona, P.J.; et al. Foamy Macrophages from Tuberculous Patients' Granulomas Constitute a Nutrient-Rich Reservoir for M. Tuberculosis Persistence. *PLoS Pathog.* **2008**, *4*, 1–14. [CrossRef]
13. Hortle, E.; Oehlers, S.H. Host-Directed Therapies Targeting the Tuberculosis Granuloma Stroma. *Pathog. Dis.* **2020**, *78*, ftaa015. [CrossRef]
14. Qamar, W.; Ahamad, S.R.; Ali, R.; Khan, M.R.; Al-Ghadeer, A.R. Metabolomic Analysis of Lung Epithelial Secretions in Rats: An Investigation of Bronchoalveolar Lavage Fluid by GC-MS and FT-IR. *Exp. Lung Res.* **2014**, *40*, 460–466. [CrossRef]
15. Stöckel, S.; Kirchhoff, J.; Neugebauer, U.; Rösch, P.; Popp, J. The Application of Raman Spectroscopy for the Detection and Identification of Microorganisms. *J. Raman Spectrosc.* **2016**, *47*, 89–109. [CrossRef]
16. Wang, H.; Zhou, Y.; Jiang, X.; Sun, B.; Zhu, Y.; Wang, H.; Su, Y.; He, Y. Simultaneous Capture, Detection, and Inactivation of Bacteria as Enabled by a Surface-Enhanced Raman Scattering Multifunctional Chip. *Angew. Chem.* **2015**, *127*, 5221–5225. [CrossRef]
17. Pahlow, S.; Meisel, S.; Cialla-May, D.; Weber, K.; Rösch, P.; Popp, J. Isolation and Identification of Bacteria by Means of Raman Spectroscopy. *Adv. Drug Deliv. Rev.* **2015**, *89*, 105–120. [CrossRef] [PubMed]
18. Schuster, K.C.; Mertens, F.; Gapes, J.R. FTIR Spectroscopy Applied to Bacterial Cells as a Novel Method for Monitoring Complex Biotechnological Processes. *Vib. Spectrosc.* **1999**, *19*, 467–477. [CrossRef]
19. Corte, L.; Tiecco, M.; Roscini, L.; Germani, R.; Cardinali, G. FTIR Analysis of the Metabolomic Stress Response Induced by N-Alkyltropinium Bromide Surfactants in the Yeasts Saccharomyces Cerevisiae and Candida Albicans. *Colloids Surf. B Biointerfaces* **2014**, *116*, 761–771. [CrossRef] [PubMed]
20. Zlotnikov, I.D.; Kudryashova, E.V. Spectroscopy Approach for Highly—Efficient Screening of Lectin—Ligand Interactions in Application for Mannose Receptor and Molecular Containers for Antibacterial Drugs. *Pharmaceuticals* **2022**, *15*, 625. [CrossRef] [PubMed]
21. Movasaghi, Z.; Rehman, S.; Rehman, I.U. Fourier Transform Infrared (FTIR) Spectroscopy of Biological Tissues. *Appl. Spectrosc. Rev.* **2008**, *43*, 134–179. [CrossRef]
22. Simonova, D.; Karamancheva, I. Application of Fourier Transform Infrared Spectroscopy for Tumor Diagnosis. *Biotechnol. Biotechnol. Equip.* **2013**, *27*, 4200–4207. [CrossRef]
23. Alvarez-Ordóñez, A.; Mouwen, D.J.M.; López, M.; Prieto, M. Fourier Transform Infrared Spectroscopy as a Tool to Characterize Molecular Composition and Stress Response in Foodborne Pathogenic Bacteria. *J. Microbiol. Methods* **2011**, *84*, 369–378. [CrossRef]
24. Novais, Â.; Freitas, A.R.; Rodrigues, C.; Peixe, L. Fourier Transform Infrared Spectroscopy: Unlocking Fundamentals and Prospects for Bacterial Strain Typing. *Eur. J. Clin. Microbiol. Infect. Dis.* **2019**, *38*, 427–448. [CrossRef]
25. Li, Q.B.; Xu, Z.; Zhang, N.W.; Zhang, L.; Wang, F.; Yang, L.M.; Wang, J.S.; Zhou, S.; Zhang, Y.F.; Zhou, X.S.; et al. In Vivo and in Situ Detection of Colorectal Cancer Using Fourier Transform Infrared Spectroscopy. *World J. Gastroenterol.* **2005**, *11*, 327–330. [CrossRef]
26. Falsafi, S.R.; Rostamabadi, H.; Assadpour, E.; Jafari, S.M. Morphology and Microstructural Analysis of Bioactive-Loaded Micro/Nanocarriers via Microscopy Techniques; CLSM/SEM/TEM/AFM. *Adv. Colloid Interface Sci.* **2020**, *280*, 102166. [CrossRef]
27. Ilie, M.A.; Caruntu, C.; Lupu, M.; Lixandru, D.; Tampa, M.; Georgescu, S.R.; Bastian, A.; Constantin, C.; Neagu, M.; Zurac, S.A.; et al. Current and Future Applications of Confocal Laser Scanning Microscopy Imaging in Skin Oncology (Review). *Oncol. Lett.* **2019**, *17*, 4102–4111. [CrossRef]
28. Chen, P.; Zhang, X.; Jia, L.; Prud'Homme, R.K.; Szekely, Z.; Sinko, P.J. Optimal Structural Design of Mannosylated Nanocarriers for Macrophage Targeting. *J. Control. Release* **2014**, *194*, 341–349. [CrossRef] [PubMed]
29. Sharma, R.; O'Sullivan, K.M.; Holdsworth, S.R.; Bardin, P.G.; King, P.T. Visualizing Macrophage Extracellular Traps Using Confocal Microscopy. *J. Vis. Exp.* **2017**, *2017*, 1–8. [CrossRef]
30. Białas, N.; Sokolova, V.; van der Meer, S.B.; Knuschke, T.; Ruks, T.; Klein, K.; Westendorf, A.M.; Epple, M. Bacteria (*E. Coli*) Take up Ultrasmall Gold Nanoparticles (2 Nm) as Shown by Different Optical Microscopic Techniques (CLSM, SIM, STORM). *Nano Sel.* **2022**, *3*, 1407–1420. [CrossRef]
31. Zlotnikov, I.D.; Kudryashova, E.V. Mannose Receptors of Alveolar Macrophages as a Target for the Addressed Delivery of Medicines to the Lungs. *Russ. J. Bioorg. Chem.* **2022**, *48*, 46–75. [CrossRef]

32. Hashimoto, M.; Morimoto, M.; Saimoto, H.; Shigemasa, Y.; Yanagie, H.; Eriguchi, M.; Sato, T. Gene Transfer by DNA/Mannosylated Chitosan Complexes into Mouse Peritoneal Macrophages. *Biotechnol. Lett.* **2006**, *28*, 815–821. [CrossRef] [PubMed]
33. Gao, J.Q.; Zhao, Q.Q.; Lv, T.F.; Shuai, W.P.; Zhou, J.; Tang, G.P.; Liang, W.Q.; Tabata, Y.; Hu, Y.L. Gene-Carried Chitosan-Linked-PEI Induced High Gene Transfection Efficiency with Low Toxicity and Significant Tumor-Suppressive Activity. *Int. J. Pharm.* **2010**, *387*, 286–294. [CrossRef]
34. Shen, F.; Chu, S.; Bence, A.K.; Bailey, B.; Xue, X.; Erickson, P.A.; Montrose, M.H.; Beck, W.T.; Erickson, L.C. Quantitation of Doxorubicin Uptake, Efflux, and Modulation of Multidrug Resistance (MDR) in MDR Human Cancer Cells. *J. Pharmacol. Exp. Ther.* **2008**, *324*, 95–102. [CrossRef]
35. Mehta, M.; Sharma, N.; Vyas, M.; Khurana, N.; Maurya, P.K.; Singh, H.; Andreoli de Jesus, T.P.; Dureja, H.; Chellappan, D.K. Interactions with the Macrophages: An Emerging Targeted Approach Using Novel Drug Delivery Systems in Respiratory Diseases. *Chem. Biol. Interact.* **2019**, *304*, 10–19. [CrossRef] [PubMed]
36. Yu, W.; Liu, C.; Liu, Y.; Zhang, N.; Xu, W. Mannan-Modified Solid Lipid Nanoparticles for Targeted Gene Delivery to Alveolar Macrophages. *Pharm. Res.* **2010**, *27*, 1584–1596. [CrossRef]
37. International, B.M.R. Retracted: Mannosylated Chitosan Nanoparticles for Delivery of Antisense Oligonucleotides for Macrophage Targeting. *BioMed Res. Int.* **2021**, *2021*, 1878130. [CrossRef] [PubMed]
38. Tan, Q.; He, L.; Meng, X.; Wang, W.; Pan, H.; Yin, W.; Zhu, T.; Huang, X.; Shan, H. Macrophage Biomimetic Nanocarriers for Anti-Inflammation and Targeted Antiviral Treatment in COVID-19. *J. Nanobiotechnol.* **2021**, *19*, 1–16. [CrossRef]
39. Lopukhov, A.V.; Yang, Z.; Haney, M.J.; Bronich, T.K.; Sokolsky-Papkov, M.; Batrakova, E.V.; Klyachko, N.L.; Kabanov, A.V. Mannosylated Cationic Copolymers for Gene Delivery to Macrophages. *Macromol. Biosci.* **2021**, *21*, 1–14. [CrossRef] [PubMed]
40. Freichels, H.; Wagner, M.; Okwieka, P.; Meyer, R.G.; Mailänder, V.; Landfester, K.; Musyanovych, A. (Oligo)Mannose Functionalized Hydroxyethyl Starch Nanocapsules: En Route to Drug Delivery Systems with Targeting Properties. *J. Mater. Chem. B* **2013**, *1*, 4338–4348. [CrossRef]
41. Ruan, G.X.; Chen, Y.Z.; Yao, X.L.; Du, A.; Tang, G.P.; Shen, Y.Q.; Tabata, Y.; Gao, J.Q. Macrophage Mannose Receptor-Specific Gene Delivery Vehicle for Macrophage Engineering. *Acta Biomater.* **2014**, *10*, 1847–1855. [CrossRef]
42. Li, C. Recent Advances in Macrophage-Mediated Drug Delivery Systems. *Int. J. Nanomed.* **2021**, *16*, 2703–2714.
43. Filatova, L.Y.; Klyachko, N.L.; Kudryashova, E.V. Targeted Delivery of Anti-Tuberculosis Drugs to Macrophages: Targeting Mannose Receptors. *Russ. Chem. Rev.* **2018**, *87*, 374–391. [CrossRef]
44. Fiani, M.L.; Barreca, V.; Sargiacomo, M.; Ferrantelli, F.; Manfredi, F.; Federico, M. Exploiting Manipulated Small Extracellular Vesicles to Subvert Immunosuppression at the Tumor Microenvironment through Mannose Receptor/CD206 Targeting. *Int. J. Mol. Sci.* **2020**, *21*, 6318. [CrossRef]
45. Jiang, H.L.; Kim, Y.K.; Arote, R.; Jere, D.; Quan, J.S.; Yu, J.H.; Choi, Y.J.; Nah, J.W.; Cho, M.H.; Cho, C.S. Mannosylated Chitosan-Graft-Polyethylenimine as a Gene Carrier for Raw 264.7 Cell Targeting. *Int. J. Pharm.* **2009**, *375*, 133–139. [CrossRef]
46. Nimje, N.; Agarwal, A.; Saraogi, G.K.; Lariya, N.; Rai, G.; Agrawal, H.; Agrawal, G.P. Mannosylated Nanoparticulate Carriers of Rifabutin for Alveolar Targeting. *J. Drug Target.* **2009**, *17*, 777–787. [CrossRef]
47. D'Addio, S.M.; Baldassano, S.; Shi, L.; Cheung, L.; Adamson, D.H.; Bruzek, M.; Anthony, J.E.; Laskin, D.L.; Sinko, P.J.; Prud'Homme, R.K. Optimization of Cell Receptor-Specific Targeting through Multivalent Surface Decoration of Polymeric Nanocarriers. *J. Control. Release* **2013**, *168*, 41–49. [CrossRef] [PubMed]
48. Zlotnikov, I.D.; Vanichkin, D.A.; Kudryashova, E.V. Methods for Determining the Parameters of Receptor-Ligand Interactions on the Model of Concanavalin A and Mannosylated Chitosans Promising Carriers for Drug Delivery to Alveolar Macrophages. *Biotekhnologiya* **2021**, *37*, 28–40. [CrossRef]
49. Zlotnikov, I.D.; Kudryashova, E.V. Computer Simulation of the Receptor–Ligand Interactions of Mannose Receptor CD206 in Comparison with the Lectin Concanavalin A Model. *Biochemistry* **2022**, *87*, 54–69. [CrossRef]
50. Zlotnikov, I.D.; Vigovskiy, M.A.; Davydova, M.P.; Danilov, M.R.; Dyachkova, U.D.; Grigorieva, O.A.; Kudryashova, E.V. Mannosylated Systems for Targeted Delivery of Antibacterial Drugs to Activated Macrophages. *Int. J. Mol. Sci.* **2022**, *23*, 16144. [CrossRef] [PubMed]
51. Mahor, S.; Dash, B.C.; O'Connor, S.; Pandit, A. Mannosylated Polyethyleneimine-Hyaluronan Nanohybrids for Targeted Gene Delivery to Macrophage-like Cell Lines. *Bioconjug. Chem.* **2012**, *23*, 1138–1148. [CrossRef]
52. Díaz-Moscoso, A.; Guilloteau, N.; Bienvenu, C.; Méndez-Ardoy, A.; Jiménez Blanco, J.L.; Benito, J.M.; Le Gourriérec, L.; Di Giorgio, C.; Vierling, P.; Defaye, J.; et al. Mannosyl-Coated Nanocomplexes from Amphiphilic Cyclodextrins and PDNA for Site-Specific Gene Delivery. *Biomaterials* **2011**, *32*, 7263–7273. [CrossRef]
53. Tiwari, S.; Chaturvedi, A.P.; Tripathi, Y.B.; Mishra, B. Macrophage-Specific Targeting of Isoniazid through Mannosylated Gelatin Microspheres. *AAPS PharmSciTech* **2011**, *12*, 900–908. [CrossRef] [PubMed]
54. Rodríguez-Lavado, J.; De La Mata, M.; Jiménez-Blanco, J.L.; García-Moreno, M.I.; Benito, J.M.; Díaz-Quintana, A.; Sánchez-Alcázar, J.A.; Higaki, K.; Nanba, E.; Ohno, K.; et al. Targeted Delivery of Pharmacological Chaperones for Gaucher Disease to Macrophages by a Mannosylated Cyclodextrin Carrier. *Org. Biomol. Chem.* **2014**, *12*, 2289–2301. [CrossRef]
55. Jain, N.K.; Mishra, V.; Mehra, N.K. Targeted Drug Delivery to Macrophages. *Expert Opin. Drug Deliv.* **2013**, *10*, 353–367. [CrossRef]
56. Azad, A.K.; Rajaram, M.V.S.; Schlesinger, L.S. Exploitation of the Macrophage Mannose Receptor (CD206) in Infectious Disease Diagnostics and Therapeutics. *J. Cytol. Mol. Biol.* **2014**, *1*, 1–10. [CrossRef]

57. Nabekura, T.; Yamaki, T.; Hiroi, T.; Ueno, K.; Kitagawa, S. Inhibition of Anticancer Drug Efflux Transporter P-Glycoprotein by Rosemary Phytochemicals. *Pharmacol. Res.* **2010**, *61*, 259–263. [CrossRef] [PubMed]
58. Van Bambeke, F.; Balzi, E.; Tulkens, P.M. Antibiotic Efflux Pumps. *Biochem. Pharmacol.* **2000**, *60*, 457–470. [CrossRef] [PubMed]
59. Macêdo, N.S.; Silveira, Z.D.S.; Patrícia, P.; Cordeiro, M.; Douglas, H.; Coutinho, M.; Pinto, J.; Júnior, S.; José, L.; Júnior, Q.; et al. Inhibition of Staphylococcus Aureus Efflux Pump by O-Eugenol and Its Toxicity in Drosophila Melanogaster Animal Model. *BioMed Res. Int.* **2022**, *2022*, 1440996. [CrossRef] [PubMed]
60. Li, X.Z.; Nikaido, H. Efflux-Mediated Drug Resistance in Bacteria. *Drugs* **2004**, *64*, 159–204. [CrossRef]
61. Bohr, A.; Nascimento, T.L.; Harmankaya, N.; Weisser, J.J.; Wang, Y.; Grohganz, H.; Rades, T.; Löbmann, K. Efflux Inhibitor Bicalutamide Increases Oral Bioavailability of the Poorly Soluble Efflux Substrate Docetaxel in Co-Amorphous Anti-Cancer Combination Therapy. *Molecules* **2019**, *24*, 266. [CrossRef]
62. Goldstein, B.P. Resistance to Rifampicin: A Review. *J. Antibiot.* **2014**, *67*, 625–630. [CrossRef]
63. Zlotnikov, I.D.; Belogurova, N.G.; Krylov, S.S.; Semenova, M.N.; Semenov, V.V.; Kudryashova, E.V. Plant Alkylbenzenes and Terpenoids in the Form of Cyclodextrin Inclusion Complexes as Antibacterial Agents and Levofloxacin Synergists. *Pharmaceuticals* **2022**, *15*, 861. [CrossRef] [PubMed]
64. Le-Deygen, I.M.; Skuredina, A.A.; Kudryashova, E.V. Experimental Methods to Study the Mechanisms of Interaction of Lipid Membranes with Low-Molecular-Weight Drugs. *Russ. J. Bioorg. Chem.* **2020**, *46*, 480–497. [CrossRef]
65. Skuredina, A.A.; Tychinina, A.S.; Le-Deygen, I.M.; Golyshev, S.A.; Belogurova, N.G.; Kudryashova, E.V. The Formation of Quasi-Regular Polymeric Network of Cross-Linked Sulfobutyl Ether Derivative of β-Cyclodextrin Synthesized with Moxifloxacin as a Template. *React. Funct. Polym.* **2021**, *159*, 104811. [CrossRef]
66. Skuredina, A.A.; Le-Deygen, I.M.; Belogurova, N.G.; Kudryashova, E.V. Effect of Cross-Linking on the Inclusion Complex Formation of Derivatized β-Cyclodextrins with Small-Molecule Drug Moxifloxacin. *Carbohydr. Res.* **2020**, *498*, 108183. [CrossRef]
67. East, L.; Isacke, C.M. The Mannose Receptor Family. *Biochim. Biophys. Acta-Gen. Subj.* **2002**, *1572*, 364–386. [CrossRef]
68. Taylor, P.R.; Martinez-Pomares, L.; Stacey, M.; Lin, H.H.; Brown, G.D.; Gordon, S. Macrophage Receptors and Immune Recognition. *Annu. Rev. Immunol.* **2005**, *23*, 901–944. [CrossRef] [PubMed]
69. Feinberg, H.; Jégouzo, S.A.F.; Lasanajak, Y.; Smith, D.F.; Drickamer, K.; Weis, W.I.; Taylor, M.E. Structural Analysis of Carbohydrate Binding by the Macrophage Mannose Receptor CD206. *J. Biol. Chem.* **2021**, *296*, 100368. [CrossRef] [PubMed]
70. Zlotnikov, I.D.; Streltsov, D.A.; Belogurova, N.G.; Kudryashova, E.V. Chitosan or Cyclodextrin Grafted with Oleic Acid Self-Assemble into Stabilized Polymeric Micelles with Potential of Drug Carriers. *Life* **2023**, *13*, 446. [CrossRef]


**Disclaimer/Publisher's Note:** The statements, opinions and data contained in all publications are solely those of the individual author(s) and contributor(s) and not of MDPI and/or the editor(s). MDPI and/or the editor(s) disclaim responsibility for any injury to people or property resulting from any ideas, methods, instructions or products referred to in the content.

*diagnostics*

MDPI

*Article*

# Assessment of Blood Microcirculation Changes after COVID-19 Using Wearable Laser Doppler Flowmetry

**Elena V. Zharkikh [1,*], Yulia I. Loktionova [1], Andrey A. Fedorovich [1,2], Alexander Y. Gorshkov [2] and Andrey V. Dunaev [1]**

[1] Research and Development Center of Biomedical Photonics, Orel State University, Komsomolskaya 95, Orel 302026, Russia
[2] National Medical Research Center for Therapy and Preventive Medicine of the Ministry of Healthcare of the Russian Federation, Petroverigsky 10, Moscow 101990, Russia
* Correspondence: ev.zharkikh@gmail.com

**Abstract:** The present work is focused on the study of changes in microcirculation parameters in patients who have undergone COVID-19 by means of wearable laser Doppler flowmetry (LDF) devices. The microcirculatory system is known to play a key role in the pathogenesis of COVID-19, and its disorders manifest themselves long after the patient has recovered. In the present work, microcirculatory changes were studied in dynamics on one patient for 10 days before his disease and 26 days after his recovery, and data from the group of patients undergoing rehabilitation after COVID-19 were compared with the data from a control group. A system consisting of several wearable laser Doppler flowmetry analysers was used for the studies. The patients were found to have reduced cutaneous perfusion and changes in the amplitude–frequency pattern of the LDF signal. The obtained data confirm that microcirculatory bed dysfunction is present in patients for a long period after the recovery from COVID-19.

**Keywords:** laser doppler flowmetry; wearable blood perfusion sensors; COVID-19; SARS-CoV-2; rehabilitation; blood perfusion; blood flow oscillations; wavelet analysis

**Citation:** Zharkikh, E.V.; Loktionova, Y.I.; Fedorovich, A.A.; Gorshkov, A.Y.; Dunaev, A.V. Assessment of Blood Microcirculation Changes after COVID-19 Using Wearable Laser Doppler Flowmetry. *Diagnostics* **2023**, *13*, 920. https://doi.org/10.3390/diagnostics13050920

Academic Editor: Juan Aguirre

Received: 30 December 2022
Revised: 20 February 2023
Accepted: 24 February 2023
Published: 1 March 2023

**Copyright:** © 2023 by the authors. Licensee MDPI, Basel, Switzerland. This article is an open access article distributed under the terms and conditions of the Creative Commons Attribution (CC BY) license (https://creativecommons.org/licenses/by/4.0/).

## 1. Introduction

The propagation of coronavirus infection, also known as COVID-19, has caused a huge number of illnesses and deaths. To date, there have been more than 650 million confirmed cases of SARS-CoV-2 infection and more than 6 million deaths worldwide (according to the Johns Hopkins University Coronavirus Resource Center). Three years after the first reported cases of SARS-CoV-2 infection, the pandemic is still far from being over. Despite the development and widespread implementation of vaccines and containment measures, COVID-19 still has a significant impact on the lives of millions of people worldwide. Emerging evidence suggests a close link between severe clinical COVID-19 and an increased risk of its vascular complications, such as thromboembolism [1]. Approximately 40–45% of cases are asymptomatic with SARS-CoV-2, but clinical observations suggest that complications may occur even in the asymptomatic course of the disease [2].

Although COVID-19 was originally considered a respiratory disease, it has now been established that it affects multiple organs and systems, including the cardiovascular system, gastrointestinal system, brain, kidney, liver, skeletal muscle, and skin of infected patients [3,4]. Recently, there is increasing evidence of the negative impact of this disease on the microcirculatory system of the blood [5–7]. It is known that SARS-CoV-2 affects the microcirculatory bed, causing edema and damage to endothelial cells, affects the development of microthrombosis, and capillary blockage, and causes a variety of other negative effects [8]. The development of these disorders, in addition to the direct threat to the patient's life and health, can also be a key factor in the development of long-term consequences of coronavirus infection, significantly reducing the quality of life of patients.

Serious concerns are caused by the fact that proinflammatory status and procoagulation activity can remain in patients for a long time after the recovery [9].

Recent observations show that a fairly large proportion of patients who have recovered from a coronavirus infection subsequently suffer long-term effects of the disease [10]. These include symptoms such as weakness, breathlessness, chest, and joint pain, confusion, memory and concentration problems (so-called "brain fog"), mood changes, etc. These and other symptoms can persist for months after the disease itself and significantly reduce patients' quality of life [11]. These disorders are referred to as "long COVID" or post-COVID syndrome. Current research is largely focused on the acute stage of SARS-CoV-2, but ongoing monitoring of the long-term effects of the disease is also necessary. In this context, the need for research into the rehabilitation of patients after coronavirus infection is clear.

There is a significant body of evidence suggesting that cardiovascular complications of coronavirus can also occur in an asymptomatic course [2], making it even more difficult to detect such complications at an early stage. This means that there will be an urgent need for both diagnostic and rehabilitative measures in the next few years for patients who have suffered from this disease. In addition, there are risks of a similar clinical outcome not only with COVID-19 but also with possible future epidemics of respiratory infections. Existing diagnostic methods routinely used in clinical practice do not allow adequate assessment of blood flow at the microcirculatory level. Currently, there is a need to develop new approaches to the diagnosis of microcirculatory disorders occurring in coronavirus infection, as well as to develop strategies for individual therapy and rehabilitation of patients after COVID-19.

Despite the widespread prevalence of the disease and the incidence of cardiovascular complications, as well as the proven extensive involvement of microvasculature in pathological processes, only very few papers have been published to date on the noninvasive assessment of blood microcirculation after COVID-19 [12–14].

One of the most common and applicable methods for diagnosing the state of the blood microcirculation system is laser Doppler flowmetry (LDF) [15,16]. This method is widely used in the diagnosis of complications of diabetes mellitus [17,18], rheumatic diseases [19], hypertension [20] and a number of other socially important diseases. Over the years, different modifications of the conventional laser Doppler technique had been introduced, including several attempts at developing wearable devices [21–23].

In the COVID-19 clinic, the main focus of research using LDF was on studying the dynamic characteristics of blood flow, including the application of functional tests. It has been shown that, during the acute phase of COVID-19, patients demonstrate a reduced vasodilatory response to local heating and reduced microvascular reactivity [24]. The correlations between microcirculatory parameters measured by LDF and laboratory test results of patients during the acute period of the disease were also analysed [25]. Another study using laser speckle contrast imaging technology demonstrated reduced vasodilation in patients with COVID-19 in response to acetylcholine and sodium nitroprusside, which persists for at least 3 months after the disease [26]. We did not find any studies in the English-language literature devoted to spectral analysis of LDF recordings in patients who underwent COVID-19. Since it is known that such analysis provides valuable diagnostic information about the state of systems regulating blood flow, including the nervous system and endothelial function, the present work aimed to fill the gaps in this area.

In this context, this work aimed to comprehensively examine the changes in blood microcirculation that occur both in the acute period of the COVID-19 disease and in the long term during rehabilitation procedures.

## 2. Materials and Methods

### 2.1. Experimental Equipment

A distributed system consisting of 4 wireless wearable microcirculatory blood flow analysers implementing LDF method "LAZMA PF" (LAZMA Ltd, Russia; in EU/UK this device made by Aston Medical Technology Ltd., UK as "FED-1b") was used for data recording in this study [27–29]. These analysers use VCSEL die chips (850 nm, 1.4 mW/3.5 mA, Philips, The Netherlands) as a single-mode radiation source. The analysers are implemented without optical fibres with direct skin irradiation from a window at the back of the instrument. This allows for avoiding fibre coupling losses as well as decreasing the movement artefacts which are common in fibre-based LDF monitors. The devices operate autonomously on internal battery power and transfer the measured signal via Bluetooth and/or Wi-Fi. The devices also have built-in motion and temperature sensors to eliminate the possible influence of motion artefacts and temperature changes on the recorded signal. When processing motion sensor data, recordings simultaneous with the subject's movements are identified as potential sources of distortion of the LDF gram and filtered using special software. The appearance of the analysers (left) as well as the options for mounting them on the volunteer's hands (right) are shown in Figure 1.

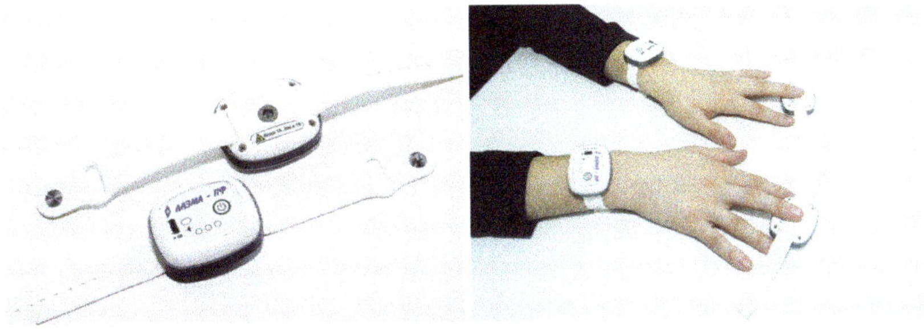

**Figure 1.** The appearance of the analysers (**left**) and the options for mounting them on the volunteer's hands (**right**).

*2.2. Experimental Protocol*

The present study comprised 2 phases. The first stage involved a dynamic assessment of the processes occurring in the blood microcirculatory system during the acute period of coronavirus infection. During routine daily LDF measurements, an 18-year-old male patient was found to be accidentally infected with SARS-CoV-2 (confirmed by PCR analysis of nasopharyngeal swabs). The patient had not been vaccinated against COVID-19 prior to the study nor did he have previous experience with COVID-19. The measurements were carried out in the supine position, each lasting for 10 min. To record signals, analysers were attached to the pads of the third fingers and big toes, as well as on the dorsal surfaces of the wrists and the inner parts of the upper third of the shins. The positioning and attachment of wearable devices on the patient's body during the study are shown in Figure 2. The measurements were taken 10 days before the onset of the disease and during 26 days after the recovery. No measurements were taken during the acute phase of the disease (7 days) because of the patient's poor well-being. A total of more than 170 LDF signals were measured and processed over the entire study period for this patient.

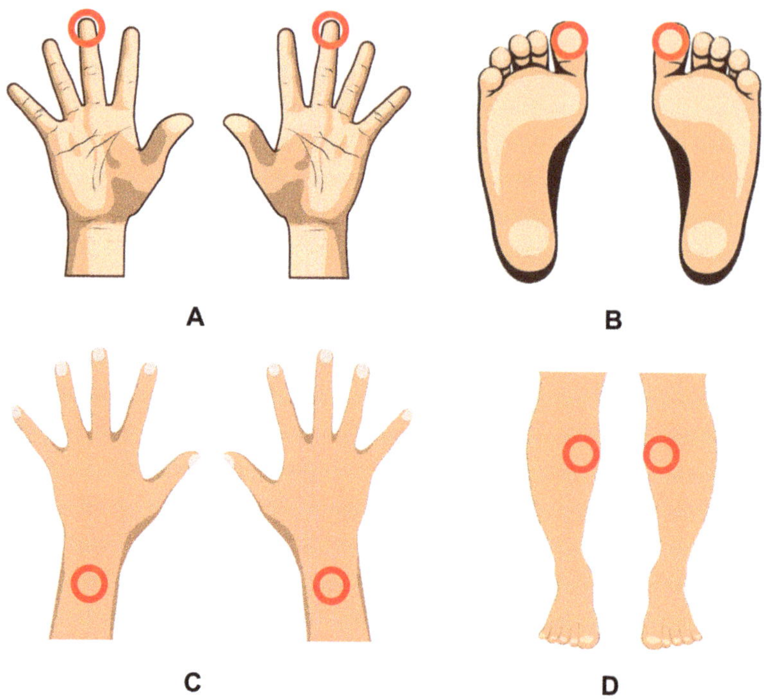


**Figure 2.** Location of the analysers on the patient's body during the study: on fingers (**A**), toes (**B**), wrists (**C**) and shins (**D**). The device attachment positions are indicated by red areas.

The second phase of the study involved the comparison of blood microcirculation parameters measured by LDF in a group of patients undergoing rehabilitation procedures after COVID-19 and a group of conditionally healthy volunteers with no previous history of coronavirus infection. The main group consisted of 23 subjects who had long COVID symptoms for a prolonged period of time after the recovery from an acute coronavirus infection and were undergoing rehabilitation in a private healthcare facility. Three of them had had a severe COVID-19 infection; all the other patients experienced moderate symptoms of COVID-19. Patients in the main group were measured between 1 and 6 months after the recovery. The mean age of the main group was $58 \pm 9$ years. The control group included 13 conventionally healthy volunteers of a matching age who were measured in 2019 before the pandemic spread, suggesting that the volunteers in the control group had never encountered COVID-19. Volunteers with any history of cardiovascular or other serious chronic diseases affecting the circulatory system were excluded from the study. The study was conducted with the subject in the supine position in a relaxed state and consisted of a 10-min measurement of microcirculation using a wearable LDF device ("LAZMA PF"). The analysers were attached to the dorsal surface of the forearms at a point 2 cm above the styloid process and on the inside of the upper third of the shins (see Figure 2C,D) as these points proved to be the most informative from the previous stage of the study. Figure 3 shows a diagram of the experimental design of the study.

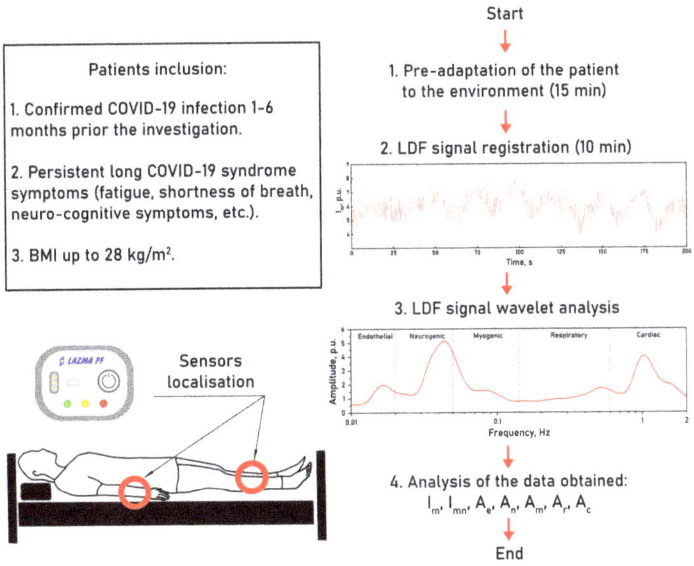


**Figure 3.** A diagram of the experimental design.

### 2.3. Data Analysis

In the present study, the analysed parameters were the value of the index of blood microcirculation—$I_m$ and amplitudes of blood flow oscillations in the different frequency bands corresponding to different mechanisms of microcirculatory blood flow regulation, measured in relative perfusion units (p.u.) [30]. The endothelial ($A_e$) band (0.005–0.021 Hz) reflects the vascular tone regulation due to the endothelium activity, both NO-dependent and independent; the neurogenic ($A_n$) band (0.021–0.052 Hz) represents the influence of neural innervation on blood flow; the myogenic ($A_m$) band (0.052–0.145 Hz) corresponds to vascular smooth muscle activity; and respiratory ($A_r$) and cardiac ($A_c$) bands (0.145–0.6 Hz and 0.6–2 Hz, respectively) carry information about the influence of heart rate and movement of the thorax on the peripheral blood flow [31,32]. To calculate the amplitude–frequency spectra of the LDF signal, we used a mathematical apparatus of wavelet transform implemented in the software of wireless wearable analysers "LAZMA-PF". This software performs a continuous wavelet transform using the complex-valued Morlet wavelet as the analysing wavelet.

In addition, the parameter of nutritive blood flow ($I_{mn}$), estimated by a well-known algorithm [33], was calculated. The use of this parameter makes it possible to estimate the distribution of blood flow along capillary and shunt vessels.

The statistical analysis of the data was performed in Origin Pro 2021 software. Due to the limited sample size, a non-parametric Mann–Whitney U test was used to check the statistical significance of differences. Values of $p < 0.05$ were considered significant. The results are presented as the mean ± SD unless otherwise indicated.

## 3. Results

The first phase of the study demonstrated that COVID-19 results in changes in microcirculatory blood flow regulation mechanisms, which can be measured by assessing the spectral characteristics of the LDF signal. The results of the measurements are shown in Table 1.

Table 1. Results of the first part of the study.

| Parameter | Wrists before | Wrists after | Shins before | Shins after | Fingers before | Fingers after | Toes before | Toes after |
|---|---|---|---|---|---|---|---|---|
| $I_m$, p.u. | $4.85 \pm 0.71$ | $4.92 \pm 0.61$ | $8.60 \pm 1.29$ | $8.25 \pm 1.61$ | $21.83 \pm 1.22$ | $21.41 \pm 1.88$ | $19.43 \pm 3.92$ | $17.48 \pm 2.78$ |
| $I_{mn}$, p.u. | $1.99 \pm 0.44$ | $1.78 \pm 0.33$ | $3.10 \pm 0.78$ | $3.14 \pm 0.78$ | $9.56 \pm 1.64$ | $8.70 \pm 1.69$ | $9.15 \pm 2.99$ | $9.71 \pm 1.78$ |
| $A_e$, p.u. | $0.15 \pm 0.04$ | $0.22 \pm 0.10$ | $0.22 \pm 0.06$ | $0.28 \pm 0.12$ | $0.56 \pm 0.37$ | $0.58 \pm 0.26$ | $0.75 \pm 0.37$ | $0.90 \pm 0.26$ |
| $A_n$, p.u. | $0.20 \pm 0.03$ | $0.24 \pm 0.11$ | $0.24 \pm 0.07$ | $0.26 \pm 0.08$ | $0.61 \pm 0.36$ | $0.64 \pm 0.32$ | $0.81 \pm 0.44$ | $1.03 \pm 0.38$ |
| $A_m$, p.u. | $0.26 \pm 0.03$ | $0.21 \pm 0.04$ * | $0.25 \pm 0.05$ | $0.24 \pm 0.05$ | $0.59 \pm 0.22$ | $0.51 \pm 0.09$ | $0.90 \pm 0.31$ | $1.18 \pm 0.41$ |
| $A_r$, p.u. | $0.27 \pm 0.01$ | $0.24 \pm 0.04$ | $0.32 \pm 0.01$ | $0.28 \pm 0.04$ * | $0.45 \pm 0.03$ | $0.42 \pm 0.03$ | $0.44 \pm 0.08$ | $0.43 \pm 0.05$ |
| $A_c$, p.u. | $0.42 \pm 0.03$ | $0.39 \pm 0.06$ | $0.49 \pm 0.05$ | $0.42 \pm 0.05$ * | $0.77 \pm 0.17$ | $0.69 \pm 0.16$ | $1.10 \pm 0.11$ | $0.98 \pm 0.16$ |

*—The significance of the difference between the values before and after the disease was confirmed with $p < 0.05$ according to the Mann–Whitney U test.

No significant changes were observed in fingers and toes in this measurement. However, there was a general trend towards a decrease in microcirculation after the disease, and also in the magnitude of the nutritive blood flow in the upper extremities. Figure 4 shows box plots of the amplitude of blood flow oscillations for the stages before and after the disease, measured in wrists and shins.

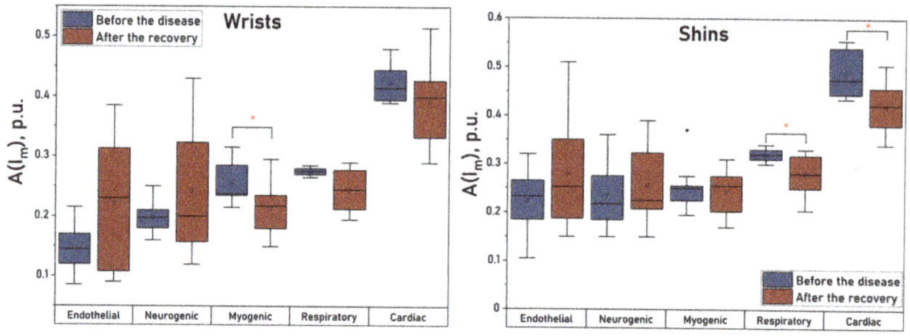


Figure 4. Box plots of blood flow oscillation amplitudes measured in wrists (**left panel**) and shins (**right panel**). *—The significance of the difference between the values was confirmed with $p < 0.05$ according to the Mann–Whitney U test.

A statistically significant decrease in the amplitude of myogenic oscillations was found in the arms after the disease. In the legs, a significant decrease in the amplitudes of respiratory and cardiac oscillations was observed. Similar changes can be traced in the upper extremities, but they do not reach statistically significant levels there. Figure 5 shows the dynamic changes in blood flow oscillations measured in wrists (a) and shins (b).

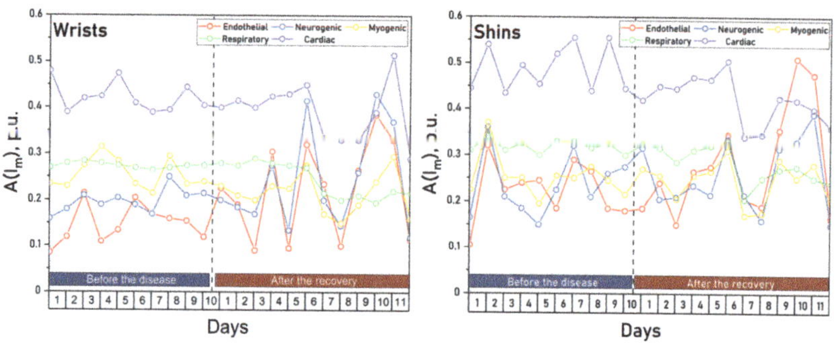


Figure 5. Changes in blood flow oscillations in the wrists (**left**) and shins (**right**) during the course of the disease and recovery.

The figures show that COVID-19 causes high-amplitude changes in the magnitude of endothelial and neurogenic blood flow oscillations immediately after the recovery, which probably caused a high variability of these values at the "After" stage and failure to achieve a statistically significant difference in them when there is a trend for their increase after the disease. These changes are especially pronounced in the upper extremities. In the legs, there is a significant drop in the amplitude of the cardiac oscillations immediately after the disease and of the respiratory oscillations one week after the recovery, which also correlates with the results obtained in the upper extremities.

The results of the second stage of the experimental study were subsequently analysed. Table 2 presents the data obtained from the second stage of the study.

**Table 2.** Results of the second part of the study.

| Parameter | Wrists CONTROL | Wrists COVID-19 | Shins CONTROL | Shins COVID-19 |
|---|---|---|---|---|
| $I_m$, p.u. | $6.70 \pm 2.13$ | $5.60 \pm 2.13$ | $8.08 \pm 2.19$ | $6.33 \pm 1.31$ * |
| $I_{mn}$, p.u. | $1.51 \pm 0.58$ | $1.15 \pm 0.48$ * | $1.93 \pm 0.99$ | $1.28 \pm 0.48$ * |
| $A_e$, p.u. | $0.20 \pm 0.11$ | $0.27 \pm 0.11$ | $0.20 \pm 0.07$ | $0.30 \pm 0.20$ |
| $A_n$, p.u. | $0.19 \pm 0.07$ | $0.28 \pm 0.12$ * | $0.24 \pm 0.08$ | $0.32 \pm 0.18$ |
| $A_m$, p.u. | $0.19 \pm 0.05$ | $0.28 \pm 0.14$ | $0.27 \pm 0.13$ | $0.27 \pm 0.12$ |
| $A_r$, p.u. | $0.20 \pm 0.03$ | $0.28 \pm 0.10$ * | $0.24 \pm 0.04$ | $0.23 \pm 0.06$ |
| $A_c$, p.u. | $0.26 \pm 0.04$ | $0.45 \pm 0.13$ * | $0.39 \pm 0.06$ | $0.54 \pm 0.31$ |

*—The significance of the difference between the values of control group and patients was confirmed with $p < 0.05$ according to the Mann–Whitney U test.

Both upper and lower extremities show significantly lower values of microcirculation and nutritive blood flow. Whisker boxes for these parameters are shown in Figure 6.

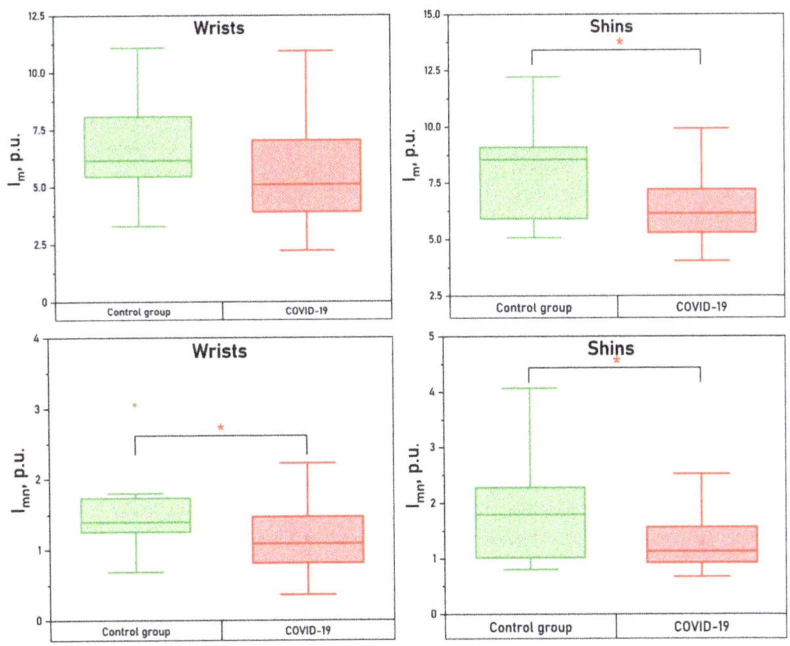


**Figure 6.** Box plots of the index of microcirculation and nutritive blood flow measured in wrists (**left panel**) and shins (**right panel**).*—The significance of the difference between the values was confirmed with $p < 0.05$ according to the Mann–Whitney U test.

An increase in overall oscillatory blood flow activity was also noted in both upper and lower extremities, with statistically significant differences in the neurogenic, respiratory and cardiac ranges in wrists. Whisker boxes for the respiratory and cardiac oscillations measured in wrists are shown in Figure 7.

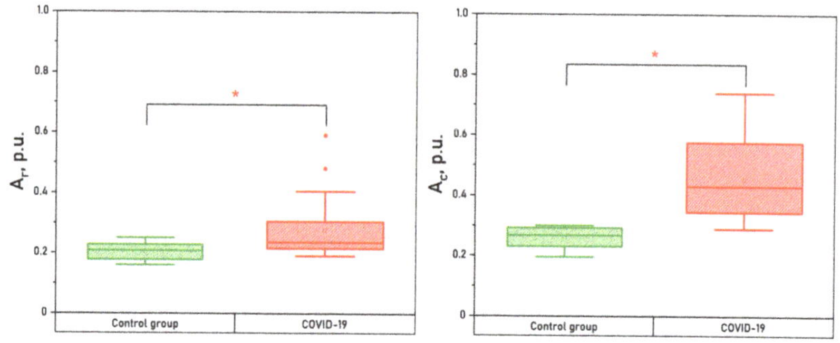


**Figure 7.** Box plots of the respiratory and cardiac oscillations measured in wrists. *—The significance of the difference between the values was confirmed with $p < 0.05$ according to the Mann–Whitney U test.

## 4. Discussion

In the present work, we obtained experimental data, which confirm the presence of microcirculatory bed dysfunction for a long period after the recovery from COVID-19. The first part of the study, which included daily measurements of one volunteer for 10 days before his disease and almost a month after the recovery, showed that after a month the parameters did not recover to their original values.

This stage of the studies revealed a decrease in the myogenic activity of microcirculation in the upper extremities. It is worth noting that the changes in the patterns of peripheral blood flow oscillations in the post-COVID phase have not yet been studied in detail. Myogenic oscillations play an important role in the process of oxygen delivery to biological tissues [34]. A decrease in myogenic oscillations leads to an increase in the dynamic resistance of microvessels and, as a consequence, to a decrease in the nutritive blood flow. Combined with the observed decrease in neurogenic regulatory activity, this change may indicate the activation of blood flow shunt pathways. In addition, some studies show that high temperature can inhibit vasomotion [35,36], so the decrease in myogenic activity revealed in our study may be a consequence of the high body temperature of the patient during the period of the disease.

The period immediately after the recovery from COVID-19 in this study was also characterized by decreased values of respiratory and cardiac microcirculatory oscillations in both upper and lower extremities (with significant differences in legs). In this case, dynamic observations show that cardiac fluctuations are reduced immediately after the disease, and respiratory fluctuations change during the week after the recovery.

Another interesting observation of this study was the increased amplitude of endothelial oscillations in the post COVID phase and the dynamics of these changes. Numerous studies demonstrate endothelial dysfunction as one of the main pathogenic mechanisms of COVID-19 [37,38], which can persist for more than 12 months after the recovery. Studies also show that long COVID-19 symptoms, especially nonrespiratory symptoms, are due to persistent endothelial dysfunction [39]. In our work, we observed increased amplitudes of these fluctuations both in the early stages of recovery from the disease and in the later stages (in the second phase of the study), although these differences did not reach a statistically significant level.

In a group of patients undergoing rehabilitation after COVID-19, the most interesting observation in the amplitude–frequency spectrum of the LDF signal, in our opinion, was an increase in the amplitude of neurogenic oscillations. A decrease in neurogenic tone leads to the dilation of the arterioles [40,41] and, consequently, the amplitude of cardiac oscillations significantly increases (which we can observe in our study).

The lumen size of skin arterio-venous anastomoses (AVA) is regulated exclusively by neurogenic mechanisms, so we can assume that they also expand amidst the decrease of neurogenic tone. The dilation of AVA leads to arterio-venous shunting of the blood bypassing the capillary channel, which explains the significant decrease of $I_{mn}$, a decrease of the number of functioning capillaries [13,14], reduction of perfusion ($I_m$) and venular overflow due to arterial blood discharge that in its turn leads to the dilation of venules [40,41] and a significant increase of the amplitude of respiratory-driven blood flow oscillations amplitude.

*Study Limitations*

The present study was conducted on a small group of patients, some of whom had comorbidities, so there is no certainty that the results will be true for the broader study population. The data obtained, however, should be taken into account for the development of new diagnostic criteria in assessing the degree of microcirculatory disturbances and rehabilitation processes in recently recovered patients. There is a need for additional studies with a larger group of patients, including patients with different courses of COVID-19 (mild, moderate, and severe disease).

Despite the already three-year history of coronavirus infection and the undoubted advantages of the LDF method for diagnosing microcirculatory disorders, there are almost no studies devoted to spectral analysis of LDF signal in COVID-19 pathology. In this pilot study, we demonstrated the possibilities of laser Doppler flowmetry coupled with the wavelet analysis of the obtained signals to detect microcirculatory disorders in patients who have undergone COVID-19 that makes it a promising tool for future research and assessment of the dynamical changes in microcirculation during the recovery process.

## 5. Conclusions

The present work demonstrates the use of laser Doppler flowmetry and peripheral blood flow oscillations analysis to diagnose vascular disorders in patients who have undergone COVID-19 in their early and advanced stages of recovery. Our work demonstrated a significant increase in the amplitude of neurogenic oscillations in the upper extremities of patients undergoing COVID-19, which, as we suggest, may be a factor preceding dilation of arterioles and venules and redirection of microcirculatory blood flow from the nutritive to the shunt pathways.

The obtained data show that optical noninvasive technologies have the potential for further application, but more research is needed to fully understand the changes in the mechanisms of blood flow regulation that occur after an infection.

**Author Contributions:** Funding acquisition, project administration, discussion and writing—review and editing, E.V.Z.; investigation, formal analysis and writing—original draft preparation, Y.I.L.; conceptualization, writing—original draft preparation and discussion, A.A.F.; methodology and formal analysis, A.Y.G.; investigation, supervision, conceptualization, and methodology, A.V.D. All authors edited the manuscript. All authors have read and agreed to the published version of the manuscript.

**Funding:** This work was supported by the Russian Science Foundation under Project No. 23-25-00522.

**Institutional Review Board Statement:** The study was conducted in accordance with the Declaration of Helsinki, and approved by the Ethics committee of Orel State University (Protocol No. 15 of 21.02.2019).

**Informed Consent Statement:** Informed consent was obtained from all subjects involved in the study.

**Data Availability Statement:** The data presented in this study are available on request from the corresponding author.

**Acknowledgments:** The authors express their acknowledgements to all the volunteers who contributed to the present study. The authors would like to thank the staff of the private treatment and prevention institution "Sanatorium Hilovo" (Pskov region, Russia) and E.V. Shuraeva for their help in organising the experimental studies.

**Conflicts of Interest:** The authors declare no conflict of interest. The funders had no role in the design of the study, data collection, analysis and interpretation, writing of the manuscript, or in the decision to publish the results.

## References

1. Levi, M.; Thachil, J.; Iba, T.; Levy, J.H. Coagulation abnormalities and thrombosis in patients with COVID-19. *Lancet Haematol.* **2020**, *7*, e438–e440. [CrossRef] [PubMed]
2. Marfella, R.; Paolisso, P.; Sardu, C.; Palomba, L.; D'Onofrio, N.; Cesaro, A.; Barbieri, M.; Rizzo, M.R.; Sasso, F.C.; Scisciola, L.; et al. SARS-COV-2 colonizes coronary thrombus and impairs heart microcirculation bed in asymptomatic SARS-CoV-2 positive subjects with acute myocardial infarction. *Crit. Care* **2021**, *25*, 217. [CrossRef] [PubMed]
3. Østergaard, L. SARS CoV-2 related microvascular damage and symptoms during and after COVID-19: Consequences of capillary transit-time changes, tissue hypoxia and inflammation. *Physiol. Rep.* **2021**, *9*, e14726. [CrossRef] [PubMed]
4. Rahban, M.; Stanek, A.; Hooshmand, A.; Khamineh, Y.; Ahi, S.; Kazim, S.N.; Ahmad, F.; Muronetz, V.; Samy Abousenna, M.; Zolghadri, S.; et al. Infection of Human Cells by SARS-CoV-2 and Molecular Overview of Gastrointestinal, Neurological, and Hepatic Problems in COVID-19 Patients. *J. Clin. Med.* **2021**, *10*, 4802. [CrossRef]
5. Gąsecka, A.; Filipiak, K.J.; Jaguszewski, M.J. Impaired microcirculation function in COVID-19 and implications for potential therapies. *Cardiol. J.* **2020**, *27*, 485–488. [CrossRef]
6. Jung, F.; Krüger-Genge, A.; Franke, R.P.; Hufert, F.; Küpper, J.H. COVID-19 and the endothelium. *Clin. Hemorheol. Microcirc.* **2020**, *75*, 7–11. [CrossRef]
7. Cenko, E.; Badimon, L.; Bugiardini, R.; Claeys, M.J.; De Luca, G.; de Wit, C.; Derumeaux, G.; Dorobantu, M.; Duncker, D.J.; Eringa, E.C.; et al. Cardiovascular disease and COVID-19: A consensus paper from the ESC working group on coronary pathophysiology & microcirculation, ESC working group on thrombosis and the association for acute CardioVascular care (ACVC), in collaboration with the European heart rhythm association (EHRA). *Cardiovasc. Res.* **2021**, *117*, 2705–2729.
8. Yelin, D.; Wirtheim, E.; Vetter, P.; Kalil, A.C.; Bruchfeld, J.; Runold, M.; Guaraldi, G.; Mussini, C.; Gudiol, C.; Pujol, M.; et al. Long-term consequences of COVID-19: Research needs. *Lancet Infect. Dis.* **2020**, *20*, 1115–1117. [CrossRef]
9. Nalbandian, A.; Sehgal, K.; Gupta, A.; Madhavan, M.V.; McGroder, C.; Stevens, J.S.; Cook, J.R.; Nordvig, A.S.; Shalev, D.; Sehrawat, T.S.; et al. Post-acute COVID-19 syndrome. *Nat. Med.* **2021**, *27*, 601–615. [CrossRef]
10. Mardani, M. Post COVID syndrome. *Arch. Clin. Infect. Dis.* **2020**, *15*, e108819. [CrossRef]
11. Raveendran, A.; Jayadevan, R.; Sashidharan, S. Long COVID: An overview. *Diabetes Metab. Syndr. Clin. Res. Rev.* **2021**, *15*, 869–875. [CrossRef] [PubMed]
12. Favaron, E.; Ince, C.; Hilty, M.P.; Ergin, B.; van der Zee, P.; Uz, Z.; Garcia, P.D.W.; Hofmaenner, D.A.; Acevedo, C.T.; van Boven, W.J.; et al. Capillary leukocytes, microaggregates, and the response to hypoxemia in the microcirculation of coronavirus disease 2019 patients. *Crit. Care Med.* **2021**, *49*, 661. [CrossRef] [PubMed]
13. Natalello, G.; De Luca, G.; Gigante, L.; Campochiaro, C.; De Lorenzis, E.; Verardi, L.; Paglionico, A.; Petricca, L.; Martone, A.M.; Calvisi, S.; et al. Nailfold capillaroscopy findings in patients with coronavirus disease 2019: Broadening the spectrum of COVID-19 microvascular involvement. *Microvasc. Res.* **2021**, *133*, 104071. [CrossRef] [PubMed]
14. Karahan, S.; Aydin, K.; Cetinkaya, A.; Sirakaya, H.A. Nailfold Videocapillaroscopy in Patients with COVID-19-associated Pneumonia in Intensive Care Units. *J. Coll. Physicians Surg.- JCPSP* **2022**, *32*, 455–460. [CrossRef]
15. Daly, S.M.; Leahy, M.J. 'Go with the flow': A review of methods and advancements in blood flow imaging. *J. Biophotonics* **2013**, *6*, 217–255. [CrossRef]
16. Martini, R.; Bagno, A. The wavelet analysis for the assessment of microvascular function with the laser Doppler fluxmetry over the last 20 years. Looking for hidden informations. *Clin. Hemorheol. Microcirc.* **2018**, *70*, 213–229. [CrossRef]
17. Zharkikh, E.; Dremin, V.; Zherebtsov, E.; Dunaev, A.; Meglinski, I. Biophotonics methods for functional monitoring of complications of diabetes mellitus. *J. Biophotonics* **2020**, *13*, e202000203. [CrossRef]
18. Mizeva, I.A.; Potapova, E.V.; Zharkikh, E.V. *Diagnostics of Functional Abnormalities in the Microcirculation System Using Laser Doppler Flowmetry*; CRC Press: Boca Raton, FL, USA, 2022; pp. 81–105. [CrossRef]
19. Mizeva, I.; Makovik, I.; Dunaev, A.; Krupatkin, A.; Meglinski, I. Analysis of skin blood microflow oscillations in patients with rheumatic diseases. *J. Biomed. Opt.* **2017**, *22*, 070501. [CrossRef]
20. Fedorovich, A.; Loktionova, Y.; Zharkikh, E.; Gorshkov, A.Y.; Korolev, A.; Dadaeva, V.; Drapkina, O.; Zherebtsov, E. Skin microcirculation in middle-aged men with newly diagnosed arterial hypertension according to remote laser Doppler flowmetry data. *Microvasc. Res.* **2022**, *144*, 104419. [CrossRef]
21. Fredriksson, I.; Larsson, M.; Strömberg, T. Model-based quantitative laser Doppler flowmetry in skin. *J. Biomed. Opt.* **2010**, *15*, 057002. [CrossRef]

22. Kozlov, I.; Zherebtsov, E.; Podmasteryev, K.; Dunaev, A. Digital laser Doppler flowmetry: Device, signal processing technique, and clinical testing. *Biomed. Eng.* **2021**, *55*, 12–16. [CrossRef]
23. Iwasaki, W.; Nogami, H.; Takeuchi, S.; Furue, M.; Higurashi, E.; Sawada, R. Detection of site-specific blood flow variation in humans during running by a wearable laser Doppler flowmeter. *Sensors* **2015**, *15*, 25507–25519. [CrossRef]
24. Sabioni, L.; De Lorenzo, A.; Lamas, C.; Muccillo, F.; Castro-Faria-Neto, H.C.; Estato, V.; Tibirica, E. Systemic microvascular endothelial dysfunction and disease severity in COVID-19 patients: Evaluation by laser Doppler perfusion monitoring and cytokine/chemokine analysis. *Microvasc. Res.* **2021**, *134*, 104119. [CrossRef] [PubMed]
25. Glazkov, A.; Ulbashev, D.; Borshchev, G.; Pulin, A.; Glazkova, P.; Kulikov, D. Skin microcirculation reactivity to local thermal hyperaemia in patients with COVID-19—A pilot observational study. *Clin. Hemorheol. Microcirc.* **2023**, *83*, 19–29. [CrossRef] [PubMed]
26. Tehrani, S.; Gille-Johnson, P. Microvascular dysfunction in patients with critical COVID-19, a pilot study. *Shock (Augusta Ga.)* **2021**, *56*, 964. [CrossRef] [PubMed]
27. Sidorov, V.; Rybakov, Y.L.; Gukasov, V.; Evtushenko, G. A System of Local Analyzers for Noninvasive Diagnostics of the General State of the Tissue Microcirculation System of Human Skin. *Biomed. Eng.* **2022**, *55*, 379–382. [CrossRef]
28. Saha, M.; Dremin, V.; Rafailov, I.; Dunaev, A.; Sokolovski, S.; Rafailov, E. Wearable laser doppler flowmetry sensor: A feasibility study with smoker and non-smoker volunteers. *Biosensors* **2020**, *10*, 201. [CrossRef]
29. Zharkikh, E.V.; Loktionova, Y.I.; Dremin, V.V.; Podmasteryev, K.V.; Sidorov, V.V.; Rafailov, E.U.; Dunaev, A.V. Comparison of wearable and bedside laser Doppler flowmetry and fluorescence spectroscopy monitors. *Opt. Technol. Biol. Med.* **2022**. *12192*, 114–117. [CrossRef]
30. Rajan, V.; Varghese, B.; van Leeuwen, T.G.; Steenbergen, W. Review of methodological developments in laser Doppler flowmetry. *Lasers Med. Sci.* **2009**, *24*, 269–283. [CrossRef]
31. Stefanovska, A.; Bracic, M.; Kvernmo, H.D. Wavelet analysis of oscillations in the peripheral blood circulation measured by laser Doppler technique. *IEEE Trans. Biomed. Eng.* **1999**, *46*, 1230–1239. [CrossRef]
32. Krupatkin, A. Cardiac and respiratory oscillations of the blood flow in microvessels of the human skin. *Hum. Physiol.* **2008**, *34*, 323–329. [CrossRef]
33. Dunaev, A.; Sidorov, V.; Krupatkin, A.; Rafailov, I.; Palmer, S.; Stewart, N.; Sokolovski, S.; Rafailov, E. Investigating tissue respiration and skin microhaemocirculation under adaptive changes and the synchronization of blood flow and oxygen saturation rhythms. *Physiol. Meas.* **2014**, *35*, 607. [CrossRef] [PubMed]
34. Aalkjær, C.; Boedtkjer, D.; Matchkov, V. Vasomotion–what is currently thought? *Acta Physiol.* **2011**, *202*, 253–269. [CrossRef] [PubMed]
35. Kastrup, J.; Bulow, J.; Lassen, N. Vasomotion in human skin before and after local heating recorded with laser Doppler flowmetry. A method for induction of vasomotion. *Int. J. Microcirc. Clin. Exp.* **1989**, *8*, 205–215.
36. Sakurai, T.; Terui, N. Effects of sympathetically induced vasomotion on tissue-capillary fluid exchange. *Am. J. Physiol.-Heart Circ. Physiol.* **2006**, *291*, H1761–H1767. [CrossRef]
37. Libby, P.; Lüscher, T. COVID-19 is, in the end, an endothelial disease. *Eur. Heart J.* **2020**, *41*, 3038–3044. [CrossRef]
38. Bonaventura, A.; Vecchié, A.; Dagna, L.; Martinod, K.; Dixon, D.L.; Van Tassell, B.W.; Dentali, F.; Montecucco, F.; Massberg, S.; Levi, M.; et al. Endothelial dysfunction and immunothrombosis as key pathogenic mechanisms in COVID-19. *Nat. Rev. Immunol.* **2021**, *21*, 319–329. [CrossRef]
39. Charfeddine, S.; Ibn Hadj Amor, H.; Jdidi, J.; Torjmen, S.; Kraiem, S.; Hammami, R.; Bahloul, A.; Kallel, N.; Moussa, N.; Touil, I.; et al. Long COVID-19 syndrome: Is it related to microcirculation and endothelial dysfunction? Insights from TUN-EndCOV study. *Front. Cardiovasc. Med.* **2021**, *8*, 1702. [CrossRef]
40. Invernizzi, A.; Pellegrini, M.; Messenio, D.; Cereda, M.; Olivieri, P.; Brambilla, A.M.; Staurenghi, G. Impending central retinal vein occlusion in a patient with coronavirus disease 2019 (COVID-19). *Ocul. Immunol. Inflamm.* **2020**, *28*, 1290–1292. [CrossRef]
41. Invernizzi, A.; Schiuma, M.; Parrulli, S.; Torre, A.; Zicarelli, F.; Colombo, V.; Marini, S.; Villella, E.; Bertoni, A.; Antinori, S.; et al. Retinal vessels modifications in acute and post-COVID-19. *Sci. Rep.* **2021**, *11*, 19373. [CrossRef]


**Disclaimer/Publisher's Note:** The statements, opinions and data contained in all publications are solely those of the individual author(s) and contributor(s) and not of MDPI and/or the editor(s). MDPI and/or the editor(s) disclaim responsibility for any injury to people or property resulting from any ideas, methods, instructions or products referred to in the content.

MDPI
St. Alban-Anlage 66
4052 Basel
Switzerland
Tel. +41 61 683 77 34
Fax +41 61 302 89 18
www.mdpi.com

*Diagnostics* Editorial Office
E-mail: diagnostics@mdpi.com
www.mdpi.com/journal/diagnostics